"With eyes wide open to the environmental problems facing the planet, elin kelsey suggests an unlikely mindset—to choose hope. *How to Be Hopeful* provides the guidance and inspiration that I so desperately need to keep doing my work."

Ginny Broadhurst,
director, Salish Sea Institute at WWU

"*How to Be Hopeful* blends scientific insight with a warm, conversational tone that invites us to courageously reimagine how to confront environmental and social crises—not through despair, but through a disciplined, evidence-informed practice of hope."

Maddy Hewitt, executive director, Near East South Asia Council of International Schools

"kelsey's nourishing work reminds us that being human and managing complex emotions is a part of the process in healing the Earth and sprouting hope around the world."

Isaias Hernandez, environmentalist and educator, @QueerBrownVegan

"As a young climate advocate,
I have spent years navigating despair in
policy rooms, protests and classrooms. *How to Be
Hopeful* offered me something rare: a framework for
choosing hope while still holding space for truth . . .
In this world that seems to be falling apart,
having hope is an act of bravery."

Lily YangLiu, youth climate negotiator
for COP29, director of the Katija Hyoungjoo
Neuber Institute, and Top 25 Women of Influence+

"*How to Be Hopeful* is a practical guidebook
that offers powerful tools and strategies for
living in strength and cultivating a peaceful mind
by acknowledging positive shifts and working
toward solutions. This book is a must-read
for anyone living with the effects of climate
injustice, which is to say, everyone!"

Rena Priest, author of *Positively Uncivilized*
and former Washington State Poet Laureate

How to Be Hopeful

elin kelsey

(and all the species that
make her life possible)

How to Be Hopeful

Empowering Practices to Overcome Despair and Act for Climate Justice

DAVID SUZUKI INSTITUTE

GREYSTONE BOOKS

Vancouver/Berkeley/London

For Esmé and Kip

Greystone Books Ltd.
greystonebooks.com

David Suzuki Institute
davidsuzukiinstitute.org

Cataloguing data available from Library and Archives Canada
ISBN 978-1-77840-262-3 (cloth)
ISBN 978-1-77840-263-0 (epub)

Editing by Paula Ayer
Proofreading by Jennifer Stewart
Cover design by Javana Boothe and DSGN Dept.
Cover illustration by bauhaus1000/iStock.com
Text design by Fiona Siu

Printed and bound in Canada on FSC® certified paper at Friesens. The FSC® label
means that materials used for the product have been responsibly sourced.

Greystone Books thanks the Canada Council for the Arts, the British Columbia
Arts Council, the Province of British Columbia through the Book Publishing Tax
Credit, and the Government of Canada for supporting our publishing activities.

Canadä

MIX
Paper | Supporting
responsible forestry
FSC
www.fsc.org   FSC® C016245

BRITISH
COLUMBIA

BRITISH COLUMBIA
ARTS COUNCIL
An agency of the Province of British Columbia

Canada Council   Conseil des arts
for the Arts   du Canada

Greystone Books gratefully acknowledges the xʷməθkʷəy̓əm (Musqueam),
Sḵwx̱wú7mesh (Squamish), and səlilwətaɬ (Tsleil-Waututh) peoples on
whose land our Vancouver head office is located.

Contents

Constellation of Practices

Preface

APPRECIATE YOU picking up this book and reading these words. Right away, I get the feeling that you are someone I would very much like to meet. I sense that because you are choosing to think about approaching the world in a way that is super difficult to do—and in a way that is urgently needed. You are choosing to tackle the overwhelming global issues we face from a hopeful orientation at a time when you are also very likely to be feeling hopeless and frustrated and betrayed and all kinds of other feelings. (After all, who picks up a book on hope when they are thriving?) It takes a lot of courage to do what you are doing. We are living in the midst of massive, multiple crises. That is a true and sad and terrifying reality. It is hard not to give yourself over to fatalistic doom and cynicism. Thank you for being brave.

It is my wish that by the time you finish reading this book, you will have found more resilience to respond to climate justice in empowered ways, and that you may actually be as impatient with narratives of doom as I am. Doomism robs us of our power to act on behalf of ourselves, each other and every other species on this glorious planet. The most powerful way to disrupt that ill-fated storyline is to learn to be hopeful. I am proud of you for having the strength not to give up your power. You need it to act on behalf of the things you love.

WHILE I AM WRITING this a crow is standing in the crosswalk on the corner. Several times the bird has attempted to get to the other side of the road while car after car drives through. No one is stopping (even though the law requires cars to stop when someone is in the crosswalk). Perhaps the drivers do not consider the crow as "someone." Perhaps they even go so far as to assume that it is the crow's responsibility to get out of the way of their car. The whole situation is tragic and upsetting to watch. Yet I also find myself in awe of the resilience of this crow. The bird is displaying such graciousness in the face of the unfairness of human-centric humans. Studies reveal that crows are remarkably good at understanding the behavior of vehicular traffic.[1] The drivers I am watching, on the other hand, are demonstrating how bad they are at considering crows.

We are living in the midst of 8.7 million other species whose resilience persists.[2] Each time we decide to show up for other species and recognize, honor and support their resilience (through nature-based solutions, for example), we are engaging in a restorative quest to heal the Earth and ourselves. The entire planetary ecosystem (and the humans and other species who depend on it to breathe and eat) thrives when we center the greater-than-human world. The resilience of other species is my personal greatest source of hope.

elin kelsey

Introduction

ARE YOU HOPEFUL?

I am asking you this question because it is at the heart of this book.

Hope is such a powerful and complex force that people have been studying it and experiencing it for as far back in history as we are able to access. And yet despite the complexities of hope, it's also a question to which I imagine you immediately know the answer. You know if you feel hopeful, just as you know if you're afraid or in love. You feel the presence of hope inside yourself.

I imagine, at this moment, your answer to this question might be "no." (After all, who else would pick up a book on *how to be hopeful*?)

Whenever I am asked if I am hopeful, the first part of my answer is always the same. "Yes," I say. "I am hopeful." The second part, however, changes all the time. It varies in response to whatever is on my mind and in my heart. "Yes, I am hopeful *and* I feel worried about _____ *and* encouraged by _____ *and* overwhelmed by _____ *and* ..." I am often surprised by what I find myself saying in the second part of my response. All kinds of thoughts and feelings emerge that I hadn't realized until someone kindly asked.

The reason the first part of my answer stays the same is that for me, hope is a choice. I wrote this book as an invitation for you to consider making the same choice. Hope is a way of being in the world that I nurture and develop no matter how bad the situation might be. I hold hope as an ongoing stance. I'm holding it in the way Quill Kukla, professor of philosophy and disability studies at Georgetown University, describes a stance as "a way of readying your body for action and worldly engagement."[1] By taking a stance, you are embodying what you believe is important. A stance is a reference point for how you approach life and your relationships with others.

The second part of my answer varies because hope is also a collection of interacting emotions, thoughts and feelings. What you or I feel, what is on our minds, changes with whatever is happening in our lives, who we are talking to, events that unfolded seconds before or years ago.

This is what I love about hope and why I think it is so tricky. Hope is a stance you can learn and practice *and* it is a collection of emotions. What I have discovered in my own experience is that holding hope as an unwavering stance forces me to actively seek out evidence of where things are moving in positive directions. This, in turn, immerses me in examples of meaningful change which inspire and motivate me. My decision to be hopeful generates the circumstances in which my feelings of hope and my agency to effect change grow.

BEFORE WE GO FURTHER, I want to emphasize that choosing hope as a stance is difficult. After all, hope is really only needed when you are confronting circumstances that make you feel hopeless. You are essentially signing on to hold hope in situations in which you will be experiencing all kinds of other feelings at the same time, including times when you feel hopeless. Many of the feelings you may be wrestling with, such as fear or worry or anger or frustration or exhaustion, may feel in conflict with hope. Within these pages we'll explore ways to develop the ability to appreciate them in concert with your stance of hope.

It's especially hard to hold onto a stance of hope because doomsaying is such a dominant cultural norm. Doom is omnipresent in everyday conversations, in the way we collectively signal what is important to us as individuals or societies, and in the media we consume 24/7. Choosing to be hopeful is thus countercultural. You will be attempting to hold a stance of hope while being constantly bombarded by a narrative of despair that is so pervasive it is rarely questioned or even acknowledged.

The thing I didn't know, when I began studying hope two decades ago, is that I was entering an ongoing conversation about how ruined and wrecked the world is that would follow me all over the planet. I knew that fear and shame had long been the way we communicate about the state of the environment, but I had not fully anticipated just how widespread and deeply embedded the narrative of fatalistic "doomism" is within societies at large.

The trouble with doomism is that it actively generates hopelessness. Hopelessness, in turn, actively disempowers you. Hopelessness is a terrible feeling. You lose your ability to trust. You can't think creatively about solutions. You feel worn down, flat, helpless. Or full of rage against the source of your pain. Hopelessness is emotionally devastating, and it also hampers your ability to address the very real challenges that must continue to be faced. Doomism diminishes your capacity to engage with the crucial issues about which you care most deeply. That is why I spend almost all of my time working to counter the culture of doomism.

This book is about hope in relation to the climate and other environmental issues, but it is also about hope more generally as a personal stance, as a way of encountering any challenge in your life, from the global to the intimate. Hope, I've come to believe, isn't a choice you make once. It's an ongoing choice you must repeatedly decide to make each time you're overwhelmed by injustice or cruelty or violation of respect or any other urgent and important issue. For me, hope is an unwavering stance that is upheld by ongoing practice. That is why this book includes other stances that support hope, as well as practices to nurture your capacity to respond from a hopeful perspective.

Don't be surprised if you are feeling uneasy as you contemplate these ideas. Considering a new approach that may not yet feel intuitive often elicits very normal feelings of skepticism or resistance. Those feelings are likely to be even more heightened because tackling urgent and important problems from the perspective of hope is probably quite different from how you (and most of us) learned to think. Bring your best critical thinking to analyze and understand what it means to hold hope and the argument for why this is a brave political stance. Allow your feelings to spark curiosity rather than falling into the trap of dismissive criticism.

I want to make it very clear that the kind of hope I am talking about is not wishful thinking or toxic positivity. I am talking about *evidence-based hope* that supports well-informed change. I feel strongly that we should expect far more from experts and educators than just an analysis of problems, and far more from leaders than denial of or commiseration about how bad things are. When you hold evidence-based hope, you demand full access to initiatives that are dismantling root causes of unjust systems. You demand up-to-date details of proven solutions and effective trends. Evidence-based hope recognizes our collective right to hear about and critically analyze not only what is broken but also what is effective in stopping or repairing that brokenness.

CLIMATE CHANGE IS a massive injustice. It is the largest, most pervasive threat to human societies, other species and planetary health the world has ever experienced. Those who did the least to cause it are forced to endure the most from it. Those already suffering from racist, unjust and unequal systems feel the consequences of climate change more profoundly. It makes day-to-day life even more precarious and difficult.

The impacts of climate change drive ecosystem collapse, biodiversity loss and extinctions, destroying the well-being and welfare of people and millions of other species. Climate change disproportionately affects younger people's lives and possible futures, both physically and psychologically. Climate change is a violation of human rights and planetary health—"the most significant intergenerational injustice of our time," according to the United Nations.[2] We are not just dealing with the urgency and enormity of climate change or the biodiversity crisis; we are dealing with these crises in an unjust and unequal world. That is why the youth climate protests happening all over the world in person and online are not calling for sustainability of the ways we currently live on Earth. They are demanding full-scale transformation to achieve climate justice.

Over the past decade, the climate crisis has become a high-level concern for the majority of people around the world. According to the Peoples' Climate Vote 2024, the world's largest standalone public opinion survey on climate change: "More than half of people globally said they were more worried about climate change now than last year, and four out of five want their countries to strengthen commitments to address climate change."[3]

Global consensus about the urgency of climate change is a powerful development. Yet in recent years, climate denial has been overtaken by a new threat: the culture of doom. Climate doomism is the belief that catastrophic warming of the planet is now inevitable and there is no ameliorative action that can be taken to avert this. It is the assertion that things are so broken, it is already too late to stop climate change. It is an unquestioned and very destructive narrative. It is so pervasive, it's likely you experience it as a simple truth.

Whenever we treat doomism as a taken-for-granted truth, we risk making ourselves complicit in creating the dead-end scenarios we most fear. As climatologist Michael E. Mann, director of the Center for Science, Sustainability, and the Media at the University of Pennsylvania, warns:

> Doomism and defeatism today pose as much of a threat to climate action as outright denial. As the impacts of climate change become ever more obvious, it's very difficult to credibly deny the problem. So, polluters have instead turned to other tactics in their efforts to block action. And among them is fanning the flames of doomism, for if we truly come to believe there is nothing we can do, then why try? That's why I focus, in my outreach efforts, on both urgency and agency—and the science

supports this. We must act now to avert catastrophic outcomes, but there is still time to act.[4]

Hope matters because our feelings about the state of the planet are based on both the very real and urgent crises we face *and* on our thoughts, beliefs and mindsets. We must reject the powerful myths of *too late, too broken* or *too big to change* that feed doomism. We deserve far more sophisticated understandings of the circumstances in which we live. We need to share what is being ruined *and* what truly works.

By actively seeking out evidence for hope, you increase your ability to hold the knowledge of how urgent and destructive the climate crisis is while at the same time rejecting the lure of doomism. Hope enables you to honor your fears and grief *and* continue to act on behalf of climate justice. When you express your well-founded fears as *fears* rather than *foregone conclusions*, you create space for meaningful action to occur.

In the face of the profound injustice of climate change, or other crucial issues impacting your life, you may be saying to yourself, "I don't want to get to a point of hope. I want my anger and my frustration to fuel me." What I would say to you is, embrace those feelings. Do not try to pacify or transform them. They are telling you that something you really care about is in danger. You do not have to choose between anger and hope. Both are powerful, activating emotions that give strength. Yet anger without hope is unsustainable. Let them coexist and work together to empower you.

Making the decision to be hopeful is a deeply personal one. Ultimately, you are the one who must decide to commit. In situations when I'm trying to decide whether to do something differently, I've found it helpful to remember that I always know a lot more about whatever I am currently doing than I know about what I am considering moving toward. In those circumstances, all I am actually deciding is whether I want to commit to discovering who I will become if I make this choice.

I REALIZE YOU MAY BE circling around these words, wondering if you can make a decision to hope before you actually believe hope is possible. Perhaps you are worrying that hope is naive. Choosing to hope is not naive. It is a pragmatic response. We must stop actively disempowering ourselves. Skepticism and cynicism are the gossamer shields we wear to protect us from the existential crisis that is hopelessness. We need evidence we can count on to overcome profound feelings of betrayal and loss of trust. That is why the *evidence* part of hope is so essential. It emboldens you to act in aid of the things you love.

The stances and practices in these pages are informed by the new science of climate emotions. They have been honed across hundreds of talks and workshops I've shared with people in many different places in the world. These stances and practices have become a way of being that feels like second nature to me, and yet, at the same time, they remain an ongoing source of contemplation, reexamination and revision.

For me, it is important that the ideas explored in this book work at both global and personal levels. I think I started this search for coherence across scale many years ago, when I met an international negotiator who was involved in conflict resolution to end sectarian violence in Northern Ireland. He had also worked in negotiations to resolve the apartheid conflict in South Africa. When I asked him what had been his most challenging negotiation, he didn't even pause. "Divorces," he said. "It is the conflicts in intimate relationships that are the most difficult to solve." He was not in any way suggesting that marital disputes are on the same plane as violent conflict. But the idea that similar, very human emotions play out at such vastly different scales stayed with me and has influenced how I approach the question of how to be hopeful. It forces me to complicate and humanize global problems in the same ways that I recognize the contradictions and intricacies of navigating personal relationships.

Perhaps it will not surprise you that I have found it hardest not to waver in holding a hopeful stance when it comes to dilemmas with people I love. So much of how we navigate the injustice of environmental crises is through acts of protest and blaming-and-shaming campaigns. Such blunt, oppositional approaches can be highly effective in these circumstances, and also utterly destructive in personal relationships. Trying to understand how best to use hope to support positive transformation across vastly different contexts and scales is compelling, rife with challenging contradictions and rewards. It's a quest that forces me to continue to look deeply at power imbalances and hidden narratives and social norms, to consider how to stand up for what is fair without alienating relationships, and to question if and when maintaining those relationships is appropriate or even possible. I'm constantly grappling with ways to hold ambiguity and complexity and compassion, how to confront my unearned privileges and to show up differently in the world, and so much more. Engaging with hope at intimate, community, national and planetary scales means I cannot disappear into the luxury of theory. Hope is something I am forever befriending and wrestling with in real situations in real time.

My desire to unleash the full capacity of hope continues to be informed by academic research, personal reflection and more podcasts, interviews and biographies than I can count. You will find a number of valuable research studies and other sources that influenced the concepts and practices in this book cited in the endnotes. I am insatiably interested in people's experiences of hope in mundane and seemingly impossible circumstances. It is my sincere wish that this book will support you in growing your own unique capacity to respond to life from a hopeful stance.

What I can tell you is that choosing to hope will change you. It will change the ways you see the world. Developing the ability to look anew—to see yourself and other species and people you love as constantly changing—will liberate you from the painful, broken cycles of fatalistic beliefs. Freeing yourself from assumptions that hold you in repetitive patterns empowers you to recognize opportunities for transformation. When you respond to doomism with evidence of a trend that is having a positive result, you increase the likelihood that a much-needed solution will flourish.

THIS BOOK FOREGROUNDS five stances that will support you in engaging with climate justice and other dilemmas you face from the perspective of evidence-based hope. I encourage you to tuck these stances deep inside your heart and call upon them daily to help you weather whatever difficult circumstances come your way. Whisper them to yourself whenever something you value turns for the worse and you're left reeling in feelings of doom. Bravely say:

Stance 1: I choose hope

Stance 2: I reject fatalism

Stance 3: I am emotional

Stance 4: I am on a reparative quest of transformation

Stance 5: I am nature

Hold these stances as active choices you make and remake whenever you need them. A number of practices are positioned throughout these pages to help you embody each stance and the hope that arises from them. You might find it helpful to create a special hope journal or to use the notes app on your phone, as many of the practices invite you to jot down responses for reflection. You can follow the book from start to finish, or begin with a stance that feels particularly difficult, and return to the specific practices that you find most helpful. Customize the practices to your own circumstances and share them bravely.

However you choose to journey through this book, I encourage you to also think about how you might create a wider culture of hope among your friends, housemates, colleagues, family or whatever community feels most appropriate. What I have discovered in talking with people over many years is that no matter how much you align with and commit to approaching the world from a hopeful stance, it is very difficult to hold onto all by yourself. The culture of cynicism and doom is just too strong to counter on your own.

Co-create cultures of support in which hope can thrive. There is something rhizomatic about collective hope. It roots and shoots like a mass of underground stems, binding us in ways we may never see. If you lose hope, the collective feeds it back, calls you back, emboldens you. This is a book about being changed and creating change together. It is a call to increase our collective capacity to respond to the complexity and urgency of the issues we face from our most empowered and creative selves. Let these stances and practices carry you through the times when hope feels effortless, and let them hearten you through the nights when the climate emergency makes being hopeful the hardest thing you can imagine.

I Choose Hope

Trees
count on storms
to move them.

With each bend and sway,
they grow "stress wood"
strong enough
to shoulder the snow
or contort their massive branches toward the light.

Trees
need adversity
to hold themselves
up.

MAKING A CONSCIOUS DECISION to engage with this troubled world from a hopeful perspective raises confronting questions. How can you choose to be hopeful in a world plagued by global crises? How can you hope when something or someone precious to you is under threat or actively being destroyed? Is it even advisable, let alone possible, to feel hopeful in the midst of profound loss? What kind of hope is capable of transforming intractable conflict? Is hope even a good idea?

It is precisely because we are living in a time of such profound injustice, urgency and uncertainty that hope is most needed. Hope serves as a psychological force or buffer that fuels your resilience in stressful and negative situations and empowers you to keep engaging with difficult issues. One of the unique qualities of hope is that it motivates you to persist in the most challenging circumstances of uncertainty, hardship and crisis, even when chances of success are slim.[1]

Hope is a multifaceted concept with deep historical roots in psychology, philosophy, behavioral science, sociology, politics, spirituality, religion, culture and more. Almost all definitions of hope, however, share two fundamental dimensions: Hope involves a desired outcome combined with an expectation, no matter how slight, that it can be attained. There is no such thing as idle hope.

Hope without agency is wishing

There are lots of things I wish for—the end of conflict, equality for other species, good health for those I hold dear. I am never short of ideas of what to wish for when the first star appears in the night sky. Wishes are things I would like to happen, with no expectation of effort or commitment on my part to do anything to make them so. Wishes are things that simply occur, as if by magic, while I am sleeping or enjoying a latte. They are desires that manifest without effort. Whenever I hear myself saying *I want this to happen but it's out of my hands* I know that I am wishing.

Hope, on the other hand, demands my involvement, my agency. One of the big barriers to hope, therefore, is believing that change is even possible. In the absence of that belief, it's difficult to convince myself to reinvest my energy and emotions *again* in seeking to make something happen that has repeatedly failed. This is especially true with seemingly intractable problems. Too often, progress on plastic pollution or the unequal distribution of wealth, for example, appears to be completely stuck. It's as if these issues are impervious to change despite concentrated efforts, often across many decades.

Believing in the possibility of change is necessary to hope. I keep a running list of ordinary "unchangeable things that are actually changing" to help myself over the hurdle of disbelief whenever it sets in. I purposely draw on unexpected examples that surprise me, to remind myself of how easily I can fall prey to making assumptions about what is stuck and what is not. Here are some of my current favorites to get you started on your own list.

Notice unchangeable things that are actually changing

1. Create a list with three columns and label them as below.

2. Add examples over time whenever you discover something in ordinary life that actually changes that you previously thought didn't or couldn't change.

A thing that appears not to change	How it is actually changing	Note to myself
Forests stand still.	Forests migrate! Most individual species of trees are rooted in one place, but forests are restless communities. They send their seeds just beyond their footprints, moving wherever the conditions to thrive are most favorable. The fossil record reveals that forests have always been on the move. Scientists now track how fast forests are moving in response to climate change.	Given that whole forests are moving without me noticing, what other changes am I missing?
My skeleton is steady like a rock.	Without noticing, I have completely replaced my skeleton about every ten years! (If I'm lucky, I'll enjoy ten full skeletons over the course of my life.) Bones are way more malleable than I ever imagined. They change in response to how we live. In 2019, scientists published evidence that more and more people now have a bony protuberance at the base of their skulls. If you have one, you may be able to feel it just above your neck. This "text neck" is occurring because of hours spent craning our necks to look at cell phones.[2]	If I can overlook something as intimate as changes in my own bones, what changes am I not seeing in a "stuck" issue I care about?

Evidence-based hope

Evidence-based hope is the art of looking for up-to-date evidence of where promise is occurring and actively engaging in meaningful ways to turn those possibilities into realities.

Evidence-based hope recognizes and rejects the culture of fear that is overwhelming us with the false belief that we are on a one-way trajectory to ruin. It empowers you to complicate the narrative, to see complexity and to amplify positive ways forward already occurring at local, national and global scales. Grounding your hope within the evidence of what is actually happening for the better helps you find courage to confront the massive issues we face as a global citizenry. As the social psychologist Erich Fromm described in *The Revolution of Hope*: Hope requires conviction about the yet-unproven.

One of the reasons the idea of hope can be off-putting to some people is that they assume holding a hopeful stance means operating in perpetual cheerfulness or some other form of toxic positivity. That is not true.

Deciding to be hopeful despite the outcome of an election or international climate change meeting or the announcement of the hottest year on record is about meeting these deeply worrying events from the most empowered position possible. Hope has nothing to do with endless positivity. Being hopeful is about being authentic. It's about choosing to show up in ways that are the most conducive to achieving meaningful results.

I recently heard a 2024 BBC Radio 4 interview with the social-political activist Gloria Steinem. I was thrilled to learn that we both share a desire to live to be a hundred, and an unshakeable commitment to the power of hope. In her ninetieth year at the time of the recording, she described herself as a "hopeaholic." She remains fully engaged in the enduring quest for equality. "We influence reality by what we expect of reality," she says. For Gloria, and for me, hope is a refusal to remain in narratives of discouragement and to firmly root in co-creating the more beautiful things that should be.

I encourage you to stop for a moment and reflect on what you are feeling. What excites you about the prospect of choosing hope? What doubts and triggers does this prospect instill in you?

Your feelings are unique to you, and I think many of us hold some beliefs about hope that can make it hard to embody. For instance, you may want to embrace hope as a stance yet feel reluctant to do so because you are holding onto the belief that hope is naive or hope is a privilege. These beliefs often come from a worthy place—you want to hold power accountable or speak out against injustice, and you fear that hope is weak and will therefore work against those aims. Challenge the voice in your head that diminishes the validity of hope. I want to encourage you to examine misperceptions you might be carrying about hope so you can actually use hope to forward your honorable intentions.

False belief 1: Hope is naive

You may fear that by choosing to hope you are letting those with political power off the hook. You may worry that hope will create a sense of complacency; that people will assume they don't have to change anything. You may be concerned that you will become misinformed and too one-sided from a positive perspective.

These fears are understandable. As Ruth Barcan at the University of Sydney describes, hope is often misguidedly equated with low intelligence or political naivety. It's as if we believe that "if you are optimistic or joyful about the world, it's because you're not bright enough to realize how corrupt it is, or not politically committed enough to jolt yourself out of your bourgeois comforts," she writes.[3] One of my friends is a senior research scientist with the Finnish Environment Institute. She recently made a personal discovery of the high value of choosing hope for her work as an ecological manager. When she shared this with her science colleagues, they dismissed it offhand. Hope, they said, is a matter for the communications team, not for scientists. For her, choosing hope means being brave enough to do so in a science culture in which that decision places her at risk of ridicule and loss of professional reputation.

Cynicism, on the other hand, is graced with the mirage of intelligence. Studies of impression management reveal that people often use cynical responses to appear more knowledgeable about a subject than they actually are.[4] It apparently works. There is a widespread tendency to mistakenly believe that cynical people are smarter. These views are further reinforced by tropes that position kind-hearted simpletons or wide-eyed idealists against those wise enough to see human nature as selfish, opportunistic and deceitful.[5]

Rejecting cynicism about the future of the planet in favor of evidence-based hope, therefore, means that you are not only choosing to go against a dominant social norm; you're also choosing to do so at the risk of being perceived to be ignorant or uninformed.

Yet you would be wise to take that risk. That's because cynics are not the masterminds they're perceived to be. Sociologists even have a name for this: the "cynical genius illusion." The reason it's an illusion is that cynical individuals generally do worse on cognitive ability and academic competency tests.[6] Whereas cynics foolishly paint everyone with the same brush, people with higher competencies perceive selfishness or greed as responses to specific circumstances, not as blanket human qualities. According to Jamil Zaki, director of the Stanford Social Neuroscience Lab: "In studies of over 200,000 individuals across thirty nations, cynics scored *less* well on tasks that measure cognitive ability, problem-solving, and mathematical skill. Cynics aren't socially sharp, either, performing worse than non-cynics at identifying liars."[7]

Many people believe being hopeful is easy. But that isn't true. Hope requires commitment and mental energy to engage with uncertain circumstances. Hope is active. It is future-oriented in seeking transformative actions.

Cynicism, on the other hand, *is* easy. It's passive. If you're already convinced something won't happen, then you don't need to do anything except sit back and watch things unfold.

So how can you overcome the cynical genius illusion? Choose healthy skepticism over blanket cynicism. The value of evidence-based hope is that it challenges you to critically analyze your assumptions. Rather than imagining that people are inherently greedy, dishonest or full of hate, evidence-based hope encourages you to gather information about who and what you can trust. As Jamil puts it: "You can be a 'hopeful skeptic,' combining a love of humanity with a precise, curious mind."

False belief 2: Hope is a privilege

Sometimes I meet people who assume that in a world shaped by crisis, only those who don't have real problems have the privilege of feeling hopeful. But hope is not a luxury. In Dharavi, one of the biggest slums in Asia, slum dwellers share a collective "politically organized hope" evidenced through the creation of an Indian network of Shack/Slum Dwellers International (sDI). Through grassroots organizing and data collection about their own living conditions, sDI groups in India (and dozens of other countries) work to have slums recognized as vibrant, resourceful and dignified communities.

"These communities oppose the politics of catastrophe, exception and emergency," writes social-cultural anthropologist Arjun Appadurai. Instead, they engage in a *politics of patience*, persistently seeking out opportunities to set precedents. They have successfully influenced planning and policy changes, increased access to water and sanitation, and implemented savings and credit schemes. Rather than "waiting for" they actively "wait to" make the next move. "As the roster of precedents grows, and the effects are multiplied, there also develops a social infrastructure that fortifies the politics of hope, and eases the period of waiting," Appadurai writes.[8] Even in the most extreme circumstances, hope exists.

Choosing hope is a commitment to believe in the capacity for change. From that standpoint, one could consider that privilege is at play whenever we step *away from* hope, and from our collective responsibility to act on behalf of necessary change.

BEING HOPEFUL IS not a naive or privileged thing. Hope is hard work. When I feel buffeted about by a chorus of cynicism, I sometimes find it helpful to connect with the physical sensation of moving myself back toward hope.

Rachel Griffiths is a participatory theater maker in the U.K. "In theater, and in organizing, we imagine and then we rehearse the hope that we want to see," she tells me at a 2024 workshop hosted by the Susanna Wesley Foundation in London. "Hope exists in the place of tension between the *now* and the *not yet*. This point of tension is very active; it is like the cat before it pounces."

Rachel kindly shared the embodied practice on the following page, which I now use to reinhabit the strength of hope. It's very simple and I encourage you to modify it (perhaps standing up or moving around) in any way that best serves you.

Embody how it feels
to move toward hope

Sit forward on a chair. Put your feet flat on the floor. Keep your back straight. Close your eyes if you wish.

Now slouch. Let your arms hang.

Imagine that you have a golden thread that goes from the base of your spine all the way up your back and out the top of your head. Take hold of the thread and, very slowly, pull your head up. When you are back in a tall seated position, let go of the string. Let yourself relax.

Now imagine a situation in which you feel truly hopeless. Find a way to express this lack of hope in your body. Where does it live? In your shoulders? In your stomach? In the tension in your neck? In the hang of your head?

Imagine you are a piece of clay. Mold yourself into an image that expresses hopelessness. Take a mental photo. What does it look like? How does it feel?

Now imagine what you will feel like when hope comes. When that situation is transformed. Move into whatever position expresses that. Slowly scan from your head to your toes to identify how you feel in different parts of your body. What emotions are arising? What is the feeling in your body of hope arriving?

Finally, go back to the position of hopelessness. Now make the journey one more time, back to your expression of hope. Notice what you did to get yourself there. Hope is a choice you will have to make over and over again. This practice can help you return to that stance and reactivate yourself in profoundly difficult moments.

Choosing hope is activating

Hope—and anger—are activating emotions. They are synergistic forces against power imbalances and systemic inequalities. "Anger often responds to disappointed hopes," explains Katie Stockdale, a philosopher at the University of Victoria, British Columbia. "We invest hope in other people to live up to the demands of morality and justice, and when they fail to do so, anger tends to ensue. But anger about injustice is also often accompanied by the formation of hopes for repair."[9]

Use your anger at injustice to propel you. Anger is very motivational because it impacts both our belief and desire systems. When you get angry, Myisha Cherry, author of *The Case for Rage*, explains in a TEDx Talk, your brain reacts in a way that makes you believe that you can influence the situation; that you can change the situation. "Chemically and biologically, when you are angry, your belief system and your desire system rise to another level," she says.[10] The result is that both your determination and your willingness to take risks on behalf of the change you desire are enhanced. Anger in support of justice is very productive and hopeful. It enables you to assert your self-respect or to demand respect for the dignity of another you hold dear. It shifts the focus away from feeling fatalistic doom toward standing up in solidarity. Solidarity requires lots of thought, talk *and* action.

Hope, like anger, fills you with a sense of agency. It enables you to look beyond seemingly intractable problems in search of more promising responses. As Paulo Freire, the Brazilian philosopher and educator, described, hope motivates us to approach the societies in which we live with curiosity, which, in turn, opens the possibility of changing the ways we live.

Hope is necessary to prompt the meaningful change you most want. First, as a way to remain motivated and resilient. Second, as a moral duty to not slip into the disempowering cycle of doomism that supports the status quo of injustice and ultimately leads to disengagement. A growing number of studies demonstrate that the more people are exposed to impactful solutions, the more hopeful and engaged they become in tackling climate risks.[11]

Holding your hopeful stance
when your heart is broken

I find it especially hard to hold onto hope when the issue is heartbreaking and I am certain that the problem has a clear solution, if only others would change.

I understand why brokenness lays such a claim on us. When there is trouble in a relationship I hold dear, I feel sick inside. Time loses dimension. I want to hide from the people I normally love to be around. If the problem comes as a shock, it can feel like an accomplishment just to keep breathing, or like I almost can't stand that it's sunny because it hurts too much to feel this bad on a beautiful day. For so many of us, caring about climate justice or species extinctions can feel much this same way. The loss and grief are visceral.

But by focusing on problems as *the* way forward, we can blind ourselves to other realities. Meaningful, positive developments may go unnoticed. Progress is not measured. Without consciously recognizing what we are doing, we end up signaling how much we care about something by chronicling the state of its demise. And yet we know, when we pause and try to remember, what two decades of positive psychology research is telling us: that the way to meaningful engagement is by focusing on strengths.

I gathered the strength to end a dysfunctional relation-ship not because of repeatedly trying to understand the problems in greater depth and detail, but ultimately because the situation was not anything like the good relationships I was lucky enough to have in other areas of my life. It was these healthy, thriving relationships that made it possible for me to take responsibility to make change, rather than all the time I had spent examining the hurtful behaviors as a route to finding a solution.

No matter whether the issue is personal, communal or global, when we are able to hold onto our stance of hope, to see what positive contributions we can add, to come from a place of confidence because others are achieving success by doing a similar thing, or to know we can count on our-selves and each other, we increase our capacity to serve the greater good.

It's easy to fall into the trap of thinking that problems are reality and solutions are pipe dreams. That's simply not true. My friend and colleague Ginny Broadhurst, director of the Salish Sea Institute, shared the following practice, which she learned from Amory Lovins, chairman emeritus of the Rocky Mountain Institute. It's a compelling way to remind yourself that problems are reality *and* solutions are reality.

Recognize problems
and solutions as reality

Draw two columns.

1. Title one column "Reality"

2. Title the other column "Also reality"

3. In the "Reality" column, list the problems that are true about an issue of concern. For example:

   a. We are deep in the climate emergency

   b. We are experiencing great losses

   c. We have wasted time

4. In the "Also reality" column, list the positive developments that are true about the issue of concern. For example:

   a. We know what the solutions are

   b. Meaningful change is happening

   c. Extraordinary transformations often occur faster than we think

5. As you work through an issue, intentionally add items to both columns. Operating in reality demands a focus on the problems *and* the solutions.

Learn to see change

For me, being hopeful is enhanced by learning to do two really important things. One is to *look for the good*. The other is to *see change*—to see it everywhere you look, whether it's in the natural world around you, in relationships you hold dear, or in the information you use to understand what's happening in the world.

The more you embrace life as ever-changing, the less often you fall into fatalism. In fact, you might be surprised by how frequently you have the opposite feeling. Even a simple walk outside can feel like a wondrous and lucky sensation. Jon McGregor captures this tantalizing feeling in his novel *If Nobody Speaks of Remarkable Things:*

> You must always look with both of your eyes and listen with both of your ears. He says this is a very big world and there are many many things you could miss if you are not careful ... there are remarkable things all the time, right in front of us, but our eyes have like the clouds over the sun and our lives are paler and poorer if we do not see them for what they are ... if nobody speaks of remarkable things, how can they be called remarkable?

Noticing the remarkable changes that surround you might even be the secret to long life—or at least the *feeling* of longevity, according to David Eagleman, a neuroscientist who teaches at Stanford University. It's common wisdom that summers felt longer when you were a kid because those sunny months occupied a larger percentage of your total young life. But according to David, that's not actually how it works. If you want the feeling that your life is longer, the trick is to slow down your perception of time. The way to do that is to collect memories. And how do you collect memories? By doing something new:

> Whatever age you are now, if you have an incredible weekend, and you look back you think, "Oh my gosh, it's been forever since I was at work on Friday," but if you have a boring weekend you think, "Oh my gosh, I was just here." And so this can happen at any age, that if you force your brain to lay down new memories, then retrospectively, that makes it seem as though more time has passed.[12]

The more you see life as ever-changing, the more likely you will be open to possibilities and opportunities. You're more likely to say "yes" to new experiences (and then figure out how to do them). And it's true, your life may feel longer and surprisingly more satisfying as a result.

Learning to see change also helps in the midst of crisis, or when much-needed progress is too slow. The more you watch for change, the more skillful you become at spotting progress happening at different time scales and in diverse contexts. This is especially useful when an issue you care a lot about takes a turn for the worse.

Progress isn't linear

Remember, progress doesn't follow a straight line. Keeping a keen eye out for the *shape* of change will help you to stay committed when the going gets tough. Remind yourself that nonlinearity is the norm of progress. I find encouragement in the words of Lee Anne Fennell, the Max Pam Professor of Law at the University of Chicago Law School:

Suppose you are organizing a rally, learning a new language, or trying to develop a downtown arts district in a city that lacks one. Things go slowly at first, and the returns seem meager. The early ralliers, the simple phrases painfully strung together, the first small gallery with limited hours and few visitors, may all have a discouraging drop-in-the-bucket quality.

But as you keep going, if you keep going, additional inputs start to bring increasing returns, and then the gains snowball for a time. Eventually, you may build a robust core of participants, a decent vocabulary, or a thriving arts district. At some later point, as success continues, things level off, and each input brings smaller and smaller marginal returns.[13]

It's an important antidote to the ways in which we talk about environmental and other climate justice losses. I frequently have conversations with people who are understandably demoralized by the approval of a pipeline or mining project or commercial development that threatens biodiversity and planetary health. "It's the final nail in the coffin," they say.

Yet when you see the shape of a particular change, you're more able to look bravely at the specific issue and to see where the trend seems to be heading. It is not the final nail. You can better weather the losses and tackle them head on. And you're less likely to get mired in defeatist beliefs, like *If it took this long for one success, it will take even longer for other successes to happen, and we won't have the stamina to endure it.* When it comes to healing, whether it be our bodies, our feelings or in the aftermath of a wildfire, change is not linear. Sometimes recovery lurches backward. Sometimes it shoots forward. Seeing change makes you better able to seize the momentum when things take a turn for the better.

Change what you measure. Healing happens too

So often, we measure damage: rates of crime in neighborhoods, numbers of endangered species, the hottest temperatures on record, the percentages of people who are unhoused. Knowing these numbers is valuable. It helps us to quantify loss, and yet it also tells us very little about how to solve these important issues.

Change that up. Be on the lookout for measurements that provide information about how things recover from damage. At the same time, work on absolutely undoing root causes that create the damage and enable it to perpetuate. I'm grateful to Cecelia Hayes, an equity and engagement manager for the government of King County, Washington, who kindly boiled down twenty-five years of experience of dismantling unjust systems into these four proven steps for system-level change:

1. Share power

2. Disrupt business as usual

3. Replace it with something better

4. Get comfortable being uncomfortable

The more we learn about respectful ways to hold the truth of injustice and how best to support resilience and healing, the better we are able to honor the people who suffer injustice, and to apply this learning to other species and ecosystems as well. That's what drew me to attend an Annie E. Casey Foundation webinar on a new scale to measure healing in people who have been harmed by community violence. I am inspired by the way this project keeps a strong focus on *both* supporting the resilience of those living in systems that create trauma *and* undoing the systems that create the problems in the first place.

"We know that many of the symptoms of trauma will persist until we end the systemic inequities that drive and exacerbate all of them. But we also know that there is something positive that can and does happen after harm," says Danielle Sered, founder and executive director of Common Justice. "We know that people in virtually all conditions heal. And so we asked, 'What if instead of measuring the absence of trauma, we measured the presence of healing?... What are the indicators of healing and can we measure them?'"[14]

Healing, she says, is about getting to the core of the harm and reclaiming possibilities. Healing isn't linear. It's multidimensional. It requires recognition and imagination. Healing is directly correlated with hope. When healing goes up, hope goes up. The scale encourages people to think about a series of statements related to pain, care and community, providing ways to recognize what healing looks like and how it feels. It inspired the following practice.

Measure the presence of healing

I encourage you to reflect on the statements below with respect to a healing journey in which you are personally engaged.

Accept the truth of the pain:
I face my past so I can heal my pain.
I endeavor to forgive myself for harm I have caused.

Nurture self and collective care:
I recognize that I deserve to heal.
I do my part so others will not go through the pain I went through.

Prioritize recovery:
I know that hurting others doesn't lessen my pain.
I choose relationships that support healing.

Embrace community:
I am connected to people who care for one another.
I use my experience to help others and myself heal.

THIS STRENGTHS-BASED APPROACH reminds me of an inspiring shift going on in the way scientists are now measuring species recoveries. For decades, the Red List created by the International Union for Conservation of Nature has been the standard for highlighting the plight of species in decline. More than 134,000 species have been graded according to how close they are to vanishing. Designations range from the most severe—extinct, or "critically endangered"—to "least concern." About a decade ago, a group of dedicated scientists gathered to ask: Is just avoiding extinction enough? Is that really a conservation success? Their response is the new Green Status of Species list.

The list accounts for the time and energy that has been invested in keeping a species from further decline, and motivates us to see what could happen in the years ahead if we continue with the efforts that have had the biggest returns. It helps identify massive accomplishments, such as species that would have gone extinct if no action had been taken. How likely is a species to make a comeback? What has to happen to make that possible? The Green Status of Species is a metric to measure how close to recovery the species is, and how close we can get in the future. It's a very hopeful and motivating approach, and a necessary complement to measuring loss.

Assume good is happening and go find it

Recognizing when things are changing for the better is not in any way saying that they're fixed, or they exist in every place they should, or that they are enough. It's simply developing the capacity to notice positive developments as they unfold.

I hold the assumption that something good is emerging around every issue I care about, no matter how bleak it appears. I make it my responsibility to find what that good is. The reason this is so powerful is that when you actively seek the good, you often find that it comes in ways that are surprising. This, in turn, increases your repertoire of possible options for responding to the issue. I don't do this to simply trick myself into being happy. That is not what it means to hope. Instead I look for good to allow myself to see the things that are possible and to bravely move beyond the immobilizing nature of doom.

The paradox of hope

When you close your eyes and imagine hope, what do you see? For me, hope often appears as openness, colorful and bright, like the sun. Recently, I met Oded Adomi Leshem, a political psychologist with the Hebrew University in Jerusalem. Oded studies the role of hope in transforming intractable conflicts. In 2017, he and Palestinian researcher Obada Shtaya (the cofounder and CEO of the Institute for Social and Economic Progress) created the Hope Map Project. This global study measures hope for peace among citizens living in conflict zones, including Israel and Palestine. In the terrible and complicated circumstances of intractable conflict, their work is to study hope where it is almost absent: amid a hundred years of strife and decades-long sorrow. "It is important not to romanticize hope," Oded says. "Sometimes hope is a burden. Sometimes it is a real challenge."

He pulls up an image on his phone of an 1886 painting entitled *Hope* by the English artist George Frederic Watts. In it hope is depicted as a woman, seated on a sinking globe, her head bowed, eyes blindfolded, her fingers plucking the single remaining string on a harp. Hope is sorrowful, fragile, bathed in despair—yet still determined to coax music from the string that remains.

This is the same image of hope that the Reverend Martin Luther King Jr. referred to in his 1959 sermon "Shattered Dreams." "Who has not had to face the agony of blasted hopes and shattered dreams?" King asks. "Of course some of us will die having not received the promise of freedom. But we must continue to move on. On the one hand we must accept the finite disappointment, but in spite of this we must maintain the infinite hope. This is the only way that we will be able to live without the fatigue of bitterness and the drain of resentment."[15]

Such is the paradox of hope. On the one hand, seeing the world as ever-changing means that we are never stuck in any given moment. Hope, in this sense, is uplifting. Contemplating the possibility of change can be transformative. On the other hand, seeing the world as ever-changing means that we never know what is coming next. We face the devastation of hopes dashed. Who knows what tragedies may be lying just around the corner?

King's words are a poignant reminder that hope transcends our individual lives. Fights for justice can take generations, even centuries, to achieve meaningful change. Many of those who fought for these causes long before us did not live to see the outcome. It is also true that accomplishments, often achieved posthumously, carry far into the future.

We are not at the starting line

When we talk about climate justice or other important issues, there is a real tendency to erase what has already been accomplished. We speak about them in the future tense. For example, *If we move away from fossil fuels, then we will...* I call this the *starting-line fallacy* because it fails to recognize the achievements that have already occurred and creates the false and discouraging impression that all the hard work lies ahead.

Support the contagious spread of hope by shifting the way you speak. Challenge the starting-line fallacy by calling attention to what is *already happening.* Emphasize how that helps us to know what we need more of:

The transition beyond fossil fuels is happening so quickly, we're now in a situation where renewables are poised to supplant fossil fuels as the world's number one form of electricity.[16] *At the same time, the collective efforts of social movements aimed at addressing inequalities and injustices have changed the global narrative from climate change to climate justice. We have proven solutions to achieving the just energy transition we need. Our challenge is to keep advocating for, amplifying and tailoring what we know works.*

By focusing on how far we have come, we overcome the pervasive belief that the issue has always been broken and will continue to be broken. We spread encouragement and perhaps even pride in the transformation we are seeking. We also gain much-needed information about what has worked to get us to where we are, which increases the likelihood of success going forward.

The agency of hope amid uncertainty

It takes a lot of courage to open yourself to even consider the *possibility* of hope when you feel overwhelmed by the state of the planet, or someone you love is seriously ill, or you have been betrayed. To hope is to commit to moving forward and embracing life with the full knowledge that you could be met with disappointment or a more profound loss. In my own life, I sometimes catch myself in the midst of an argument with someone I dearly love, being drawn toward thinking of leaving the relationship altogether. In those moments, I'd rather immerse myself in the certainty of despair than proceed with the risk of uncertain joy. The possibility that things might not work out is just that painful.

Nevertheless, an element of uncertainty must exist for hope to arise, according to philosopher Ernst Bloch. Hope never comes with guarantees. The potential for disappointment, therefore, is always with us. As political philosopher Jakob Huber writes, "in cases where the odds of making a difference are dim, we need hope in order to sustain our commitment to action."[17]

Each of us has our own tolerance for uncertainty. Yet whether we wish it were true or not, uncertainty is a normal part of life. In some circumstances, uncertainty can even be pleasurable. When the stakes are low, you might enjoy not knowing, according to evolutionary biologist David Krakauer at the Santa Fe Institute.[18] That's part of the attraction of watching sporting events or reading mystery novels. The thrill is not knowing what is going to happen and enjoying witnessing what unfolds.

Accepting that none of us can truly know what will happen next can help us look differently at far more serious situations that have become deeply entrenched. One of the things that holds violent conflicts in place is people on both sides believing that the conflict is inherently unresolvable. The certainty that no solutions can be found has a negative impact on people's willingness to work toward peacebuilding, which escalates conflict, and the cycle repeats itself.[19] In the process, the certainty of the conflict is reinforced. "The extreme duration of intractable conflicts makes people experience the conflict as a familiar and predictable—however dire—reality. Peace, on the other hand, is strange and unfamiliar for those who have known conflict all their lives and therefore entails uncertainty," explains Oded Adomi Leshem, the political psychologist I mentioned earlier. "People who are more at ease with uncertainty allow themselves to wish for peace while those that seek certainty do not dare to wish for peace as much."[20]

The relationship between hope and peace is crucially important. Empirical studies conducted in conflict zones around the world consistently demonstrate that hope is not only an outcome of the peace process but one of its sources. In the midst of the violent sectarian conflict in Northern Ireland known as the Troubles, hope was found to be positively associated with a lower desire to retaliate, and a higher motivation to forgive.[21] The more one believes peace is possible, the more one supports peacebuilding efforts. "Hope," Oded tells me, "stands out as the most robust predictor of people's willingness to act for peace."

Craving certainty is natural, and yet uncertainty is inherent to life itself. Perhaps you, like so many of us, struggle with wanting things to be permanent and knowing that nothing lasts forever. According to Dan Siegel, clinical professor of psychiatry at the UCLA School of Medicine and executive director of the Mindsight Institute, recognizing this paradox and letting both of these feelings coexist can help to ease the dread of anticipating ominous troubles, while opening the door to approach whatever unfolds with a sense of harmony, adaptability and curiosity. Learning to sit with rather than run from the unknown can open up new and necessary ways of thinking.

Uncertainty teaches us to live in the now. I remember listening to an interview with Lyse Doucet, the BBC's chief international correspondent. She has covered all of the major conflicts in the Middle East since the mid-1990s, yet she is clear that she does not wish to be labeled as a war correspondent. Instead, she wants the focus to be on the greater arc of life. "Everyone I've known who has lived in war wants nothing more than to get out of war," she said. "It's been a life of ups and downs, of light and dark, humor and happiness, and so I don't want to be defined by that horrible three letter word, war."[22]

Instead, Lyse is a strong proponent of cultivating calm in the midst of chaos, of remaining curious, of trying to live each day as your best day. "I live constantly bathed in this sense of gratitude," she said. "Gratitude for the life that I live, the friends that I have, the people who love me and I love in return."

My friend Judy Long is a chaplain who specializes in care for caregivers tending to loved ones in life-altering medical circumstances. She echoes Lyse's intentional approach to life. "There is what is happening," Judy says, "and there is how you choose to relate to it."

Recognize insecurity as a product of unjust systems—and call them out

It is important to differentiate our acceptance of uncertainty from the kind of insecurity that is generated by inequality, scarcity and the unreliability of institutions. Six out of seven people *worldwide* feel plagued by feelings of insecurity, according to a 2022 United Nations Development Programme report:

People's sense of safety and security is at a low in almost every country, including the richest countries, despite years of upwards development success. Those benefiting from some of the highest levels of good health, wealth, and education outcomes are reporting even greater anxiety than 10 years ago.[23]

When governments fail us, the consequences are not only political but psychological. As individuals, we bestow trust upon large institutions just as we do in close relationships. When they break that trust by failing to uphold their obligations, we suffer the physical and mental health anguish of heartbreak.

Astra Taylor is a filmmaker, activist and author of *The Age of Insecurity*. In a 2023 interview, she underscores how capitalism creates a manufactured insecurity that impacts all of us:

> We can recognize the differentials, that insecurity hits those who are marginalized and poorest and most oppressed the hardest. But it's also present at every rung of the income ladder, and that's part of capitalism's grip. It's why people can't get off the treadmill and say, "I've got enough." In a society with healthcare, pensions, or other forms of a safety net, you wouldn't have to be rich to be secure. But the irony of capitalism is that even rich people don't feel secure![24]

An intense and entrenched conflict is at play between those who believe the answers to the climate crisis can be found within the global system of capitalism, and those who recognize capitalist modes of production to be *the cause* of climate change. "The principle of infinite growth upon which contemporary capitalism is premised is incompatible with a planet that by its very nature has finite resources," write business accounting scholars Michele Bigoni and Sideeq Mohammed.[25]

If you find yourself feeling paralyzed by the idea that we need to undo capitalism in order to tackle climate change, or relieved that the root systemic cause is being expressed by Astra in such compelling ways, embrace your reaction. Notice if you are responding to something you assume is stuck or that is positioned in a starting-line narrative. Find ways to engage with it from a strengths-based perspective.

Perhaps, for example, you recognize the nonstop growth and consumption that defines capitalism is by definition unsustainable, and yet you also feel capitalism spurs technological innovations, such as renewable energy, that reduce greenhouse gas emissions. It is your choice whether to work for what you believe to be achievable changes within the current system, or to work toward dismantling these systems of power, or a combination of the two.

Hope requires us to imagine new possibilities and look at root causes of problems, even at systems like capitalism that may feel stubbornly embedded. Learning to see the narratives that support the systems that shape our relationships with ourselves, each other and the planet, and how to shift within and beyond them, is what empowers us. Choosing the stance of hope enables us to act anew in aid of the things we love.

I Reject
Fatalism

When my sister fell ill
that beautiful year
when we were both healthy young mothers
and then—she was not
I remember struggling without clarity
Am I helping her to live—or to die?

A wetland is never forever
a bog or a marsh or an estuary.
It is a place in eternal transition.

In the night
as I lay beneath my duvet
longing for your presence
while suffering the sorrow of your betrayal
I felt trapped beneath the weight of the in-between.

And yet,
my sister is alive
the swamp is now a river
and we are becoming
whatever it is
we become.

FIND IT REALLY HARD to be present with my fears. Pretty much all the time, but especially in situations where my fears are well-founded. There is no disputing the fact that climate change is a global tragedy that is causing horrific destruction and suffering. Fear is a healthy response to the very real and present danger of the climate crisis. We fear for our own futures, and for the futures of other people and other species.

What I have noticed, though, is how quickly I move from my justifiable fear to a position of fatalism. Fatalism is the voice that says *this problem cannot and will not be solved.* As soon as I move into the belief that what I fear will inevitably continue to happen, I am robbing myself of my agency to do anything to stop it.

I understand why I jump to fatalism or even cynicism. In most cases it's because my fears have been proven right *many* times before. My fear is being fed by a growing mountain of evidence that supports my feeling that there is nothing I can do to stop the tragedy from unfolding. I take sad comfort in the conclusion that I am powerless to alter its course.

This fatalism, however, is a dead end. It is a passive place where all I can do is be victimized by the brokenness of the situation. It is a place of chronic sorrow with no foreseeable end; a place where I can blame myself or others for putting me in this terrible predicament.

When I muster the courage to shift from fatalism back to fear, I discover I have all kinds of other emotions.

I am heartbroken or grieving or blaming or furious. I feel betrayed by someone I should have been able to count on. Though it is sometimes excruciatingly painful to look clearly at the thing I love that is being lost, it is also necessary. Sitting with my fear allows me to keep actively engaging with the issue, and to search for ways forward that honor each of the feelings I have.

Next time you feel overwhelmed by the state of the world, remind yourself you cannot alter anything, or escape the pain you are enduring, from that place of fatalism. Find the courage to be with your fear. Encourage yourself to respect your emotions. By doing so, you are acknowledging how personally challenging this predicament is. You are also reawakening your humanity by rejecting systems of patriarchy and capitalism that have conditioned you to devalue your emotions or to view them as signs of weakness.[1] Respecting your emotions makes it easier to be brave. When you feel fatalistic, encourage yourself to make the powerful choice to keep returning to fear. This is what will enable you to begin to approach the issue in new ways.

Find the courage to be with your fear in order to resist sliding into fatalism

1. Notice when you are stuck in dead-end stories. (You can recognize them because they often come in the form of blanket statements such as *No one is doing anything* or *People are too selfish*.)

2. Catch yourself whenever you start shifting from fear toward fatalism—a feeling of resignation that something is already destined to happen.

3. Try to notice when you are feeling afraid of what will happen next, or worrying about the future.

4. Accept that your fear may be well-founded. Given what is happening, you may be wise to be skeptical that anything good could possibly happen in this situation.

5. Check to see if you are too depleted or exhausted or convinced that *this situation is too broken to ever be fixed* to process further.

6. If so, give yourself a rest and lean into your communities of support. Be compassionate about how difficult this is. Say out loud: *This is hard*.

7. When you are ready, find the courage to circle back to feeling again.

CONTINUED ▶

8. Remind yourself that no matter how scary things feel in this moment, none of us ever really know what is going to happen in the future.

9. Let your fear flow through and out of you. Don't give it the power to settle into yourself as full-on fatalism.

10. Identify other feelings that reside alongside your fear (anger? sadness? worry? grief?).

11. Accept that all of these are real and helpful feelings. Don't try to talk yourself out of or justify any of them. Just feel them.

12. Remember that fear and love are deeply interwoven. Your fear is being fed by your deep concern that something you love—yourself, a relationship, the Earth—is under threat.

13. Turn your back on disempowering fatalism. Follow your thoughts and feelings toward how you could feel and act on behalf of what you love.

REJECTING FATALISM IS not an easy stance. It is perhaps one of the most difficult. It is difficult because the things that fuel our fatalism are often true and scary and profound. Rejecting fatalism is more difficult still because we are living in a media culture that systematically reproduces fatalistic doom.

You are living in systems
that make you feel doomed

The crises we face are bad enough without demoralizing ourselves. Yet that is what is happening. It is as if there is an unspoken agreement that we don't need to provide evidence that things are getting worse, because everyone knows that's just true. You know the rhetoric, and the very real feelings it elicits: *Too little, too late, no future. The world is wrecked and it's only going to get worse. We are doomed. We're fucked.*

If I could give you a magic wand, I would ask you to use it to transform climate change media. Because no matter if you're scrolling the news, checking social media or chatting with friends, almost everything you hear about climate change is presented in a narrative of doom.

Maxwell Boykoff leads the Media and Climate Change Observatory, an international, multi-university collaboration based at the University of Colorado Boulder. For twenty years the observatory has been researching trends in climate change reporting. Today, they monitor more than 120 sources across television, radio and newspapers in fifty-nine countries. What they have found is that all over the world, "climate catastrophe," "climate emergency" and other intense phrases are now standard practice in climate change reporting.[2] As Anne Saab, a professor of international law, writes, "Climate change is widely described as an impending catastrophe, a looming apocalypse and a path to irreversible damage that may even result in human extinction."[3]

At first glance, you might think this is a good thing. After all, climate change is a very real and urgent crisis, and it's important that we identify it as such. Yet a significant body of research now demonstrates that when climate change is presented within the doomsday rhetoric of apocalyptic language and images, we spiral into fear, reactivity and depression. When you're bombarded and exhausted by the "thirst for environmental drama and exaggerated rhetoric," there's a higher likelihood you will disengage, according to Mike Hulme, founding director of the Tyndall Centre for Climate Change Research.[4] Rather than looking at the complexity of the underlying systemic issues, we become mired in what sociologist David Altheide describes as a "politics of fear."[5] Narratives of doom leave people feeling overwhelmed and disengaged rather than empowered to act.

The solution is to disentangle the all-too-real problem of climate change from the cultivation of fear. If this doesn't happen, we risk making the actual crisis worse. That is the tragic result of the way COVID-19 was reported in the U.K. media, according to the abstract of a 2020 research study:

> Newspapers' editorials, headings, and articles of that period framed the coronavirus pandemic in terms of fear-mongering by dramatizing reports on the epidemic in China, by metaphorically presenting the coronavirus as a deadly living thing approaching Great Britain and finally hitting the country like a tsunami, by repeatedly emphasizing the globality of the pandemic and inadequacy of the government's measures to curb the disease.[6]

The virus was portrayed as the "nightmare of pandemics," "World War Flu," and "the ultimate threat to humanity," fear-based rhetoric that *increased* people's feelings of anxiety over and above the very significant threat of COVID-19 itself.

This is a hard idea to grasp because there is a common perception that you need to scare people into taking threats like pandemics (or climate change) seriously. Or at least that any extra anxiety is not a big deal, because the consequences of *not* taking the thing seriously enough would be so much worse. Yet adding a pile of anxiety on top of an already challenging crisis does more harm than good. "Control, fear and shame tactics don't work," according to research led by Vijay Bharti, a West Virginia University academic who studied the effects of public health communication during the COVID-19 pandemic. Instead, messaging is most effective when it emphasizes choice rather than exerting control, uses language like "can" rather than "should," and frames actions as personal decisions rather than resorting to blame and shame.[7]

Per Espen Stoknes, codirector of the BI Center for Sustainability and Energy in Norway, uses the term "apocalypse fatigue" to describe the fear, guilt and passive victimhood we experience when we are repeatedly exposed to doomsday rhetoric. You feel hopeless, which can easily give rise to cynicism, apathy and an exhausted feeling that there's really nothing you can do to change the problems that concern you.

As digital strategist Kaylee Gardner explains, "A lot of our current climate change communications are written with a scarcity mindset, focusing on how our actions deplete resources and our quality of life may be decreasing soon. Simply put, this framing of information causes panic."[8]

That panic, however, does not inspire action. It does the opposite. According to the American Psychological Association's Task Force on the Interface Between Psychology and Global Climate Change: "Attempts to create urgency about climate change by appealing to fear of disasters or health risks frequently lead to the exact opposite of the desired response: denial, paralysis, apathy, or actions that can create greater risks than the one being mitigated."[9]

Such urgent narratives may very well stem from a deep desire to make people understand the urgency of the climate crisis. Yet in addition to creating a doomist media culture that ultimately leads to disengagement, this approach is also not responsive to the fact that most people on the planet are now very aware of and worried about climate change. The idea that "if they knew better they would act better" (and that we can make them *know better* by inciting fear) is outdated and ineffective in creating the type of engagement that the climate crisis requires.

Prophesying doom hampers progress

It makes me angry when progress is made, and instead of acknowledging an important step, the media shouts, *It won't last!* COVID-19 caused the largest drop ever in carbon emissions in recorded history.[10] Yet in Canada, as in many other countries, negative prophecy headlines dominated. "Why your reduced carbon footprint from lockdown won't slow climate change," led CBC News on May 23, 2020.

Not only is this headline discouraging, it's wrong. Canadians had been reducing their carbon footprints long before the lockdowns occurred. A 2024 Environment and Climate Change Canada report states: "Between 1990 and 2022, the amount of GHGs emitted per person decreased 17% from 21.9 to 18.2 tonnes of carbon dioxide equivalent (CO_2 eq)."[11] In October 2024, the EU announced the achievement of the largest annual drop in greenhouse gas emissions in decades—emissions fell by 8 percent over 2023. "The steep drop in planet-heating pollution in 2023 is close to the fall recorded in Europe at the start of the Covid-19 pandemic, when travel restrictions grounded planes and shuttered factories," reported Ajit Niranjan in *The Guardian*.[12]

This is not to say there isn't much more work to do in Europe and in Canada. It *is* to say that you have a right to know the accurate numbers and trends rather than fears cloaked as facts. This knowledge provides further information about what is working and empowers us to take additional actions to amplify these important changes.

If you believe climate change is unstoppable, you're increasing the likelihood it will be

Narratives of doomism rely on shock and fear to motivate people to take urgent action in the hopes of averting disaster. These narratives operate on the assumption that people are unaware of the climate crisis and that they need to be scared into action.

It's a flawed strategy. Most people already know there is a problem. Unfortunately, when people are already worried, narratives of doom can become self-fulfilling prophecies. A study of fifty thousand people from forty-eight countries found that those who believe climate change is unstoppable are less likely to engage in personal behaviors or support policies to address climate change.[13] This makes sense. If you really believe you can't stop something, why would you even try?

This is what makes climate doomism so pernicious. It preys upon the very people like you who care the most. Climate doomism manifests as an existential crisis that renders too many of us unable to act on an issue we perceive as an imminent threat to our lives. The situation is a double tragedy, given that it is not too late to act on climate change, and engaging with the issue is exactly what we must continue to do to solve the crisis.

University of Maine climate scientist Jacquelyn Gill has noticed a shift from climate denial to climate doom since 2018. She is keen to clarify that the belief that "nothing can be done" just isn't true, and it actually makes things worse. Speaking to the Associated Press, she says, "I refuse to write off or write an obituary for something that's still alive. We are not through a threshold or past the threshold. There's no such thing as pass-fail when it comes to the climate crisis."[14]

Fighting climate change means rejecting doomism as well as denial and staying engaged. I greatly admire Christiana Figueres, former executive secretary of the United Nations Framework Convention on Climate Change, who continues to hold hope even after decades of confronting fraught political processes on behalf of climate justice. In November 2024 she coauthored an open letter to the UN demanding an overhaul of the failing intergovernmental process.

At the same time, she remains stubbornly optimistic about advancements in the global effort to decarbonize economies. Appearing on a January 2025 Mongabay podcast, she says:

> I think the problem here, and the one that I struggle against constantly, is a view of the world that is black and white. Either we are perfect, or we're doomed. I don't think we are either of the two … I used to think that it was our collective responsibility to guarantee to future generations that they would have a perfect world. And now that I am a recent grandmother, I really look back at

that and I go, "My God, we cannot guarantee to future generations that they're going to have a perfect world." We cannot. So, what can we do? We can do our darndest and we can wake up every morning and make a choice and say, "Where am I going to put my energy today?"[15]

I'm inspired by Christiana's refusal to accept the oversimplification of complex issues into polar-opposite categories of "perfect" or "doomed," or to expect ourselves to act as either "for" or "against." She is modeling what it means to work across the full spectrum of engagement, embracing ambiguity and nuance and long-term commitment to change that matters.

Social psychologist Jonathan Haidt might say that Christiana is one of those people who have won what he calls the cortical lottery—those who just naturally have an expansive worldview that allows them to see the good and to bathe in strong feelings of gratitude. If you are lucky, that may also be true for you. But if, like for so many of us, it isn't, it is useful to see the capacity for gratitude as a skill, as something that can be learned with practice.

Choose the parts of you that you wish to grow

I really appreciate this empowered way of thinking. Rage is in us. Hopelessness is in us. Generosity is in us. Kindness is in us. We have centers in our brains for altruism and revenge.[16] What matters is what we choose to grow within ourselves.

Part of rejecting fatalism is actively choosing to operate from a perspective of compassion and gratitude. That's because these emotions help us see beyond the narrow keyhole of self-interest and social comparison. When you foreground thankfulness and appreciation, you gain a fuller and more expansive perception of the world and yourself. "MRI studies show that until we get in touch with our compassion, we misjudge almost everyone, including ourselves, for the worse because we are so threatened," explains Fred Luskin, director of the Stanford University Forgiveness Project. "The minute we touch compassion, we start seeing them without our prejudices. We see them as people ... We are able to see with more nuance."[17]

Developing your capacity for gratitude is effective in overcoming fatalism because it shifts our attention away from dead ends and limitations and toward a heightened sense of possibilities and personal empowerment. One of the most effective ways to practice gratitude is to recognize people who are "tailwinds" in your life; people whose support or inspiration propels you forward.

Appreciate the tailwinds
that power you forward

Turn to whoever is nearby and describe one person in your life that you are grateful for and why.

That's it. That's all you have to do.

Later, if you wish (no need to add pressure, just if and when you can), send a message to the person to whom you are grateful telling them what you said. Don't worry about trying to get the words exactly right. What will shine through is the gesture; the simple fact that you took the time to do it. Think of it in the spirit of this saying generally attributed to Maya Angelou: "People will not remember what you did. They won't remember what you said, but they'll remember how you made them feel."

We underestimate how much
progress we've made

I am grateful to Thomas Gilovich, professor of psychology at Cornell University, for using the metaphor of tailwinds and headwinds to explain our human tendency to attend more to the disadvantages than the advantages in our lives. "We focus intensely on the headwinds that slow us down but take the tailwinds that propel us forward for granted," he said on a 2023 podcast of *The Hidden Brain*.[18] When it comes to climate change, we are so overwhelmed by the force of the headwinds, we underestimate the accomplishments that have been made on the journey so far.

Most people don't realize how much progress we've made on climate change, writes Dana Nuccitelli of Yale Climate Connections. As of 2023, "climate policies and clean technologies deployed over just the past eight years have already erased a full degree Celsius of global warming from the future world in 2100."[19] A livable, sustainable future for all is still within reach, if we accomplish the ambitious actions we know we must take, according to Hoesung Lee, the former chair of the Intergovernmental Panel on Climate Change.[20]

If reading this triggers worries of greenwashing, please acknowledge those feelings *and* also consider that we are in a very different position than we were not that long ago. Climate change concern is widespread around the globe. In an analysis of forty countries, on average, 69 percent of respondents stated that they consider climate change to be an extremely serious or very serious threat. International scientific consensus on human-caused climate change is at 99 percent, and climate change denial has been decreasing continuously for the past decade.[21]

We misjudge how other people feel about climate issues

A 2024 study across 125 countries demonstrates widespread demand for climate action:

> Notably, 69% of the global population expresses a willingness to contribute 1% of their personal income, 86% endorse pro-climate social norms and 89% demand intensified political action. Countries facing heightened

vulnerability to climate change show a particularly high willingness to contribute.[22]

Yet if I asked you if you believe others care about climate change and are willing to act, chances are high you might say "no." Part of what feeds cynicism is our tendency to misjudge how others behave or think: a phenomenon sociologists call *pluralistic ignorance.* In the United States, for instance, twice as many people support climate change policies and actions as those who don't. Americans believe, however, the opposite to be true. They underestimated popular support for climate policies by nearly half.[23]

Neuroscientist Todd Rose shares many examples of how *collective illusions* fuel distrust and drive political polarization in the U.S. Much, if not most, of what we believe others believe is inaccurate. According to a survey that asked people about their goals for the country, he writes, "Americans of all political stripes agree on the need to address climate change, privately ranking the issue's priority at number three ... Yet when we asked respondents where they thought other Americans would rate its importance, it came in at number thirty-three."[24]

Climate justice issues require collective action, which makes underestimating others' concern and support a big deal. You are far more likely to act if you believe other people care or are doing something. I feel encouraged to keep pressing governments for more action when I encounter research like that from the London School of Economics in late 2024, that shows climate framework laws have already had a positive impact on climate action and the way it is implemented at a national level in New Zealand, Ireland

and Germany. I am relieved to learn that climate framework laws actually work, and that sixty countries have them. And I'm thankful for the details this study provides about what effective climate laws entail and what still needs to be improved. According to Alina Averchenkova, distinguished policy fellow at the Grantham Research Institute on Climate Change and the Environment, who published the report:

> To be effective ... the laws should include key building blocks, such as a long-term net zero target, interim targets or carbon budgets, and requirements for emission reductions plans, regular reporting, assessment and independent reviews of progress. Legislators must be alive to the fact that current laws do not sufficiently provide for public participation. This should be corrected to increase a sense of national ownership and shared understanding of the benefits of the transition, which is essential to prevent social backlash against climate action.[25]

Instead of being crushed by the feeling that national governments aren't doing anything, I have a clearer sense from Alina's findings of where progress is happening in national climate law. There are specific measures I will now demand from other governments to increase the likelihood of successful climate action.

Climate change reporting is a business that impacts how you feel and act

Unfortunately you're unlikely to hear about the strong support for climate change policies or what works in terms of climate change law in the press. Despite a literature explosion of research studies on climate change in recent years, selection bias toward media coverage of catastrophic physical events means that social science insights like these go unreported.[26]

You already know, but it's worth repeating: Almost everything you think about climate change is a result of how it is reported in the media. And newsroom decisions about what climate change stories make the news are driven by many factors. News production is big business. Before the internet, judgments of "newsworthiness" were primarily reflections of the journalist's biases and interests (conscious or unconscious), the political or social orientation of the news organization, and the desires of the owners or advertisers, as well as government pressures. Now, in the digital age, all those influences still exist, and are further complicated by the algorithms used by digital platforms which ultimately determine what gets published.

According to Muhammad Khamaiseh, an editor at Al Jazeera Media Institute, a fierce debate rages between two sides in the news business. There are some who argue that a story is "newsworthy" based on its capacity to draw attention to what the public needs to focus on for its own best interests—in other words, what genuinely impacts the public interest. And there are others for whom "what determines the newsworthiness of an event is its potential for going 'viral,' provoking debate and trending, potentially causing journalists to overlook important yet 'boring' events."[27]

The result is that only a slim slice of climate change research makes it to the news. Just 2 percent of 51,230 research articles published in 2020 on climate change received extensive media attention, and almost all of those focused on heat waves, floods, storms and other meteorological events. "Studies of shrinking glaciers and melting sea ice are very popular... and papers on megadrought, the longer lifetimes of hurricanes in a warming world," say the authors of a 2023 research report on climate change media.[28]

The trouble is, this approach works *against* engaging you in climate action. Research reveals that these images draw attention, but they also leave us feeling powerless and hopeless, with a diminished sense that we can do anything about these issues and a weaker intention to act. And, when we reduce climate change to catastrophic physical events, we neglect our human responsibility for the systems of racism, inequality and injustice that underpin and aggravate these dire situations.[29]

This narrow view of climate change is further aggravated by the fact that a significant portion of the research featured in climate change reporting originates in the Global North, and more specifically, the United States. "More than 40% of news coverage of climate change research originated from studies in just six high-profile journals—three from the Nature portfolio and two published by the American Association for the Advancement of Science, along with the *Proceedings of the National Academy of Sciences of the United States of America*," write climate change researcher Marie-Elodie Perga and her colleagues in a 2024 issue of *Eos*.[30]

On top of all of this is the negativity bias that dominates media reporting in general. People often worry that if they focus on solutions-oriented content they will be engaging in biased news consumption. Yet they are already, often unknowingly, doing exactly that. The portrayal of reality in the news is *negatively* biased. Writing in a 2022 research article, Dutch communication scholars Toni G. L. A. van der Meer and Michael Hameleers state:

In their overly negative and event-driven reporting, news media are commonly found to portray a biased reality. Overall, good news is seen as synonymous with no news as negativity is believed to be more attention-grabbing and therefore garners the highest audience ratings and number of clicks. In this predominant focus on negativity, sensational and exceptional events often gain disproportionate news attention. The infrequency of incidents paradoxically explains the level of news coverage and therewith isolated and negative events are mistakenly presented as daily reality. Media's tendency to present the negative and exceptional as reality can cultivate a distorted worldview among its audiences. For example, negatively biased coverage can induce irrational fear perceptions, lower general well-being, and increase support for radical right-wing political parties. Such worldviews are at odds with the principles of a well-functioning democracy.[31]

These issues are so problematic, even the World Association of News Publishers is now asking: "Is journalism inadvertently contributing to climate inaction?" To stoke climate action, we need a transformation in climate change reporting. We need a shift in focus from "newsworthiness" to "news usefulness." We need climate change news that "balances urgency… with efficacy."[32] We need proven approaches that both renew our faith in the solvability of social crises and give us more information about what works. We demand this by choosing to reject fatalism.

Healthy media consumption
increases your capacity for action

Given the detrimental impacts of doomist headlines, part of holding a stance to reject fatalism is to consciously make choices about news consumption. Luckily, research on news media literacy interventions—ways to help you think more critically about the information you are consuming—reveals that learning about how news media is skewed toward the negative (through, for example, negativity bias and clickbait) increases your ability to avoid spirals of cynicism and gain more control of your news diet.

You embolden yourself—and strengthen our collective power as citizens—when you expect more than the same old story of climate change as a looming catastrophe that people don't care about: a story that implies the only chance of averting certain disaster rests solely in the hands of national governments that are failing us, and that all you can do is passively watch as the horror unfolds.

Each time you reject headlines that frame climate change conferences as the "last chance before catastrophe strikes," you are taking a stand against reporting that conceals the reality of suffering that is already taking place.

When you realize you are operating in a system of discouragement, you might notice yourself becoming impatient with such lopsided reporting. Climate change solutions are happening all over the world. But you're unlikely to hear about them. According to a 2022 analysis of the headlines from 336,000 climate change articles on Google News, less than 1 percent mention climate change solutions.[33]

How often do you begin your morning by checking the news? It's a familiar, widespread habit. No doubt you do so for good reasons—to stay informed, to bear witness in solidarity against atrocities, to hold politicians and others accountable and more.

Yet each time you consume news, you're making a decision that profoundly affects you and others. What you see matters. A decade ago, researchers revealed that "individuals who watched just three minutes of negative news in the morning had a whopping 27% greater likelihood of reporting their day as unhappy six to eight hours later."[34] In the ensuing years, many studies have revealed that the personal impact of news is far greater than you might imagine. It affects our dreams and increases our risk of heart attack, anxiety and depression. It makes our individual worries seem worse. It causes us to overestimate risk and to stereotype other people.[35]

If you're under the age of twenty-five you're likely to access your content via social media. The emotional fallout and psychological distress of seeing and hearing the horror of war, mass shootings, forest fires, floods and other disasters in real time, 24/7, via YouTube, Instagram and other platforms increases as sophisticated algorithms deliver more gruesome images.

Despite your concern for the suffering, research reveals that seeing this graphic content does not motivate collective public action to resolve tragedy. Instead, the amount and type of this content and its biases create more mental distress and poorer functioning.[36] Looking at graphic bad news doesn't keep you informed, and it doesn't contribute to solving the problem. Instead, what emerges is collective trauma that disempowers you and other viewers from engaging with the issue.

To this, it might feel instinctive to say something like: "Well, it makes sense that it puts me in a worse mood, but ultimately that's because things are just bad, and feeling bad is the price of paying attention to the news." I urge you to challenge this assumption. Feeling bad doesn't mean you are being informed. Doomscrolling evokes existential anxiety. It can foster pessimism about human nature. Indeed, researchers at the University of London found that brains triggered by doomscrolling exhibit effects that resemble those in medical studies where scientists have temporarily turned optimists into pessimists by suppressing the activity of the inferior frontal gyrus—the area of the brain that actively filters bad news as it acquires new information.[37]

My editor, Paula Ayer, noticed that when she sees something horrible online, instead of stopping to feel sad or angry about it, she often keeps scrolling, perhaps in an attempt to self-soothe by putting some new stimulation into her brain. She worries that the mix of gossip, jokes, celebrity news and horror all swirling around together, as is often the case on social media, further causes desensitization to awful things that shouldn't be happening.

Demand more balanced news

To hold your stance of hope, I encourage you to consciously demand news that empowers rather than harms. I often draw strength from the wisdom of Eve Tuck, founding director of the Provostial Center for Collaborative Indigenous Research With Communities and Lands at NYU. From her, I have learned to question the damage I inflict when I participate in systems that position me as a spectator to other people's agony. Ask yourself: *Am I engaging in "disaster voyeurism" when I focus on the crisis unfolding on my screen?*

As Susan Sontag wrote in *Regarding the Pain of Others,* "Perhaps the only people with the right to look at images of suffering of... extreme order are those who could do something to alleviate it... or those who could learn from it. The rest of us are voyeurs, whether or not we mean to be."

I've made a personal decision not to watch moments of horrific suffering that disempower me and re-victimize others, but instead to inform myself through podcasts and other longer-form media that centers the voices of the individuals and communities most affected.

When I immerse myself in more complicated stories of despair and resilience, situated within an understanding of the global systems and histories that fuel a crisis, the people involved become real. I find the courage to really look at the horror of what is happening. I see the generosity that sits alongside evil. I am better informed and able to throw my strength behind those working on behalf of the greater good.

This is the strength of solutions journalism. When journalists rigorously investigate and report on what people are doing to solve an issue, our sense of agency and empowerment rises. We see the full complexity of the dilemma and the solutions.

Solutions journalism supports social change by illuminating outdated narratives and rigorously investigating societal responses that are working to solve ubiquitous issues. I'm better able, for example, to catch myself from falling into an outdated narrative of "the poor" because I've been following the work of Zoe Greenberg, a reporter at *The Philadelphia Inquirer.* Her stories complicate the narrative, exposing the inequality and systemic racism that create conditions in which some people suffer from poverty, while others have stakes in maintaining the systems that perpetuate that injustice.[38]

Narratives hold the status quo in place. They influence what we know, feel and do as individuals and as whole societies. That is why your capacity to hope will grow each time you're able to spot a narrative and to challenge its power. The more you develop your capacity to see something as a widely shared story rather than a forever truth, the better you'll be able to also see promising counternarratives that may better support the greater good. Think of this like the story of "The Emperor's New Clothes." Notice how often you see what you have been told to see—and how quickly what you see changes when someone points out the obvious flaw in the story.

Reporting on how Mexico City averted an all-out drought in the summer of 2024, Maya Averbuch resisted the sensational "Day Zero of water" storyline that was commonly reported in other media. (Day Zero refers to the critical point at which a city's water supply is predicted to be completely depleted, leaving communities in crisis.) Instead, she revealed that even in a normal year, 80 percent of low-income residents with water pipes don't receive water daily. Rather than framing the issue as a one-off catastrophe, she portrayed water management in Mexico City as an ongoing problem that demands long-term solutions. Maya quotes Rodrigo Gutiérrez Rivas, a researcher focused on constitutional and water rights at the National Autonomous University of Mexico:

> The solutions aren't simple. You have to resolve the problems of housing and water at the same time, to construct a city that's for everyone and not just for the rich...

Water became one of the main issues during the [2024 local elections] and now that it's won by a wide margin, the government of Mexico City has a huge opportunity to transform the model.[39]

Perhaps you still feel that negative news is necessary to hold those in power accountable. Yet studies show you are more likely to stop watching the news and thus step away from your political responsibilities when you're overwhelmed by doom. Four out of ten people *globally* say they sometimes or often avoid the news, according to a 2022 study by the Reuters Institute. This is a big problem for democracy, given that "news avoidance is associated with less political knowledge, less political participation, and greater misbeliefs."[40]

Solutions-oriented news, on the other hand, enables you to stay informed because you are not experiencing the same damaging psychological and physiological costs.[41] It is also powerful at driving change because it showcases meaningful actions that are already happening, and thus equips us to point to what others have and demand that it exist in this circumstance too.

So how do you take the reality of events seriously, *and* reject the onslaught of negativity that pushes you to avoid the news? What empowering choice can you make for yourself and the issues you care about when you wake up tomorrow?

You're already taking the most important step of thinking about the impacts of the media that you consume. The next step is to expect the news you consume to empower you. You have a right to hear about solutions that work. Immerse yourself in solutions-oriented content.

Solutions journalism

Solutions journalism is a rapidly growing field of journalism, which means there are new sources emerging all the time. You can start with the Solutions Journalism Network and their searchable database, or simply Google "solutions journalism" and "climate justice" (or whatever issue you're interested in). Many news outlets have a solutions-oriented column, or weekly newsletters you can get sent directly to your device. If it's helpful, take a peek at my favorite solutions-oriented content providers on my website. You might also wish to check out sources of peace journalism (such as Peace News Network) and constructive journalism. Choose to find and follow reporters (and news sources) who foreground what is producing meaningful results. Claim your right to hear fuller stories about climate change and other crucial world events.

Some environmental journalists, too, find focusing on solutions enables them to keep going year after year. "There is no sense in being hopeless until there is nothing left to hope for," writes Carol Linnitt, editor-in-chief of *The Narwhal*. "Big, bad stories are all over the headlines, but less mentioned are how hard people are working to protect the natural world every day... There is simply too much at stake to give up."[42]

Source solutions-focused news that empowers not harms

1. Spend the first 15 minutes of whatever time you normally engage with the news *only* looking at solutions-oriented content.

2. Make a quick entry on your phone (or whatever journal-keeping format you regularly use) to record how the solutions-oriented content makes you feel.

3. Repeat these steps every day for 10 days.

4. Notice what solutions inspire you. What ways can you further engage with them or amplify their impact?

5. Share solutions journalism sources across your broader community.

Here are some reflections from people who have engaged in this practice, to give you a sense of what you might experience:

> I feel energized, despite the time. It's 11:11 p.m. as I type this. I think this sudden rise in energy levels can be attributed to the hopeful narrative present in both news articles, which still feels unfamiliar to me. I have begun noticing that positive emotions tend to arise when I start reading about something from a solutions-oriented lens. For example, right now I feel enlightened and eager to learn more about what I just finished reading. It is comforting to know that I can allow myself to

feel empowered simply by being mindful of which news sources I engage with regularly.

I do find it challenging to read about these solutions and not critique their attempts to make things better. University has taught me to apply a critical lens to problems, solutions, advertisements, papers. I have realized how harmful my mindset is for filtering content. Always looking at the negatives is a hard system to shift. This practice is helping me make that shift.

I always feel inspired seeing how many people are working on climate solutions, especially when the initiatives they are working on support local communities. A lot of climate solutions that I read about are technology and industry related and I appreciate reading about nature-based and community-based solutions. It helps to remind me that there are a huge number of people engaged in a huge number of different solutions. I often feel like we are at the starting point of tackling climate change, but I need to remember that there is so much good work already going on.

If you're a teacher, allocate at least
half of all class time to solutions

This is precisely what my former graduate student Graeme Mitchell, the cofounder of the Institute for Global Solutions at Claremont Secondary School in British Columbia, is doing. The decision to restructure the curriculum came in response to Graeme's discovery that students in his tenth-grade class had consumed *46,000 hours* of social media in a single school year. Like all good educators, Graeme engaged the students in a collective study to identify what they were consuming and the impact it was having. What did they find? They were drowning in an avalanche of sensationalist doom. As Graeme describes in his 2024 TEDx Talk, the students had little knowledge about real-world solutions and scored poorly on tests about advances in health care, renewable energy and other positive global trends.[43] And they were really afraid. Indeed, the students were suffering from what Anthony Leiserowitz, director of the Yale Program on Climate Change Communication, calls a "hope gap": They were deeply concerned about what's happening on the planet *and* they felt powerless to do anything about it.[44]

Changing *half* the content the students consumed in class to solutions-oriented material is helping to shift the culture toward evidence-based hope. Today, the students are involved in a frenzy of solutions-based community projects, scoring higher in their knowledge about world events and reporting more optimistic mindsets.

(For more strategies for teaching about climate change, check out the Special Appendix for Educators on page 242.)

Reject fatalism by celebrating
that you are constantly, gloriously wrong

"It is difficult to learn what you already know." I really love this quote. It impresses me that it is nearly two thousand years old and that it was first spoken by Epictetus, a Greek slave who later gained fame as a sage and mentor to Emperor Hadrian in Rome. His wisdom has not only stood the test of time; it has influenced contemporary theories and practices of cognitive behavioral therapy, the current gold standard of psychotherapy.

I also value the cautionary tone of the quote. It prompts me to try to recognize whatever I'm sure I know—and then to question if it is still true. It's quite a tricky thing to do, because my own personal taken-for-granted ideas rest in my mind and slip out of my mouth without my even registering they are there. They are embedded in my language and my habits. Like the air I breathe, they are simply the familiar and comfortable facts by which I live my life.

Yet when I do catch myself, and take the time to check, I recognize how frequently my ideas have been fed by racist, speciesist and colonial narratives I grew up within. It's my responsibility to reflect on and actively unlearn these narratives. I do this by challenging my "facts."

All kinds of beliefs I'm carting around turn out to have expired. Recognizing this reality is a powerful way to reject fatalism. Like a forgotten jar of pesto hidden behind the milk in the back of the fridge, these "facts" are well past their sell-by date and are no use to anyone. But unless I actually go looking for them and replace them, they linger, contaminating whatever is near at hand without my even noticing.

The practice on the following page, "Check the expiry date on what you are sure you know," is one of my favorites. It's as simple as realizing that the last time you looked at an issue was a few months back and then using your best solutions journalism sources to see what is actually happening now. Remember, because environmental content is so heavily weighted toward problems, you'll need to use positive keyword searches—and likely wade through a lot of gloomy headlines—before you'll find the positive trends.

I give lots of talks throughout the year and before each one, I go through and update every slide. I check the expiry date on dozens of current examples of climate justice and environmental trends that are achieving results we need to amplify. I need to be thorough because so much is happening, it's almost impossible to stay up to date.

Check the expiry date on what you are sure you know

1. Listen for blanket statements, whether they come from you or someone else. (It's helpful to me that my former mother-in-law used to collect "sweeping generalizations"—broad claims that don't recognize the complexity or specificity of what's actually happening.) In conversations about climate change, listen for assumptions, such as "No one cares" or "No one is doing anything about..." or "Renewables are too expensive."

2. Assume that whatever the blanket statement is, it is likely out of date.

3. Go looking for timely evidence to help you determine the current state of play.

4. Remember to assess timeliness: not only by how recently the information was gathered, but also by whose voices are being centered. Current information acquired through a lens that doesn't include diverse perspectives and intersectional outlooks, for example, is past its expiry date.

5. Find the courage to share what you are finding, especially when you land in conversations where tired old blanket statements are being treated as taken-for-granted truths.

Witnessing change invigorates you

What you'll quickly discover by doing this practice is that everything is always changing, not only for the worse, but also, quite often, for the better. Staying up to date has made me much more aware of the scope and scale and details of what is happening around the world in response to the climate justice crisis. What I've noticed is how eager I am to share what I have discovered—the first voyage of a fully electric ten-thousand-ton container ship in China; the rise of "climate quitting" and other ways people are pushing their current and potential employers to prioritize environmental and social justice values; how the repair of road culverts in Washington state restored access to 571.18 miles (919.22 km) of potential habitat for salmon and steelhead.[45] Instead of feeling the heaviness of fatalistic climate doom, I feel more fully informed about the issues *and* invigorated by solutions I didn't even know were occurring.

I've also found that checking the expiry date helps me to have more meaningful conversations with others. Not just because I'm better prepared with up-to-date content, but because the challenge of trying to stay up to date keeps me humble. People often ask me, "What do you do when you're talking to someone and you can't change their mind?" Honestly, what I have found is that once you start approaching the world with this expiry-date idea, you realize how often you are stuck in old ways of thinking yourself. It has made me more open to the idea that part of the problem is that I am also stuck. That is something I can do something about. I have agency to see where I am contributing to fatalism through an entrenched belief. It opens my eyes to situations where I am perpetuating an assumption that serves no use.

I can have empathy for someone else who is stuck, but my real power lies in double checking what I am saying, and openly, proudly, enthusiastically declaring when I discover I am wrong.

One of the reasons I started doing this practice is that people who come to a talk I'm giving sometimes share deep worries about an environmental issue for which their particular concern is simply out of date. They are grieving a specific issue that has moved in positive directions of which they are unaware. Because they are assuming the problem is always a problem, they don't think to look to see whether or not anything has changed.

I think this is often the case for things that really matter to you. You might be afraid to look because you fear finding out that the situation has actually gotten worse. But try to be brave and actually look. Because when your fears become your facts, fatalistic doomism flourishes. It is absolutely true that you may find things are even worse than you expected, and, sometimes, things get better too. Hope rises when you go looking and discover that more than a third of the world's electricity now comes from renewables, or that conservation efforts have successfully halted the decline of elephant populations in southern Africa over the last twenty-five years.[46]

In an unexpected twist, this practice has made me quite impatient with doom that is fueled by inaccurate assumptions. Climate justice and other environmental issues are real and heavy enough. I feel heartsick when our emotions are being spent on something that is no longer the worry that it once was.

It's comforting to know that almost all of us are wrong about more things than we realize. I am a huge fan of Gapminder, a Swedish foundation co-created more than twenty years ago by Hans Rosling, physician and author of the international bestseller *Factfulness.* The focus of the foundation is to identify systematic misconceptions about important global trends, or, as they put it: "to fight devastating ignorance with a fact-based worldview everyone can understand."[47] They use big data analysis to reveal global trends, which is great, but what I really love is that they also track how many of us are wrong about those trends. It's motivating to learn, for instance, that the annual number of oil spills from tankers worldwide has decreased tenfold since the 1970s. I didn't know that. It's also strangely comforting to realize that 64 percent of people surveyed didn't know this. Most of us don't know that oil spills have become rarer. Knowing that is motivating, because we can learn what we collectively did to make that positive trend happen, and what we need to do to make it even more effective.

Unearthing "past-its-sell-by-date" content that has been operating as a truth helps you to keep fatalistic doomism at bay. Each example reminds you that all kinds of things that we collectively treat as unchangeable truths about how the world works often turn out to be wrong. This opens up all kinds of possibilities to navigate what you might have thought was a dead end. Checking the expiry date is hopeful because it rejects fatalistic doomism by revealing opportunities for change that weren't visible until you looked.

AS YOU WORK YOUR WAY through this book, check in with yourself. Do you have strong emotional responses to some of the content? Are there moments when you are interested in the ideas yet feel a heightened reaction? Take a moment to notice what might be happening. You may find it helpful to explore the chapters in bite-sized pieces, giving yourself time to pause and feel and think. The reason I am suggesting this is that many of the ideas in this book make sense from a thinking perspective, yet they can feel unsettling. This hearkens back to how deeply most of us have been conditioned to be wary of anything resembling a solution. The idea that the only or best way to change something wrong is to focus on what is broken is so ingrained, you may find yourself reacting each time you see a solution mentioned in the book, even while you may also be interested to read about it. Feeling and thinking are so deeply interwoven, they are constantly influencing each other. You may wish to circle back and reread earlier sections of the book to help remind you and reinforce your thinking—and feelings—about hope.

STANCE 3

I Am
Emotional

I am too heartbroken to go outside
and be with the warmth of the sun
Today I am a
dark inside girl.

NOTHING STAYS STUCK FOREVER. That is what I try to remind myself when the too-heavy-to-raise-my-body-from-the-couch, curtains-drawn, everything-is-broken despair hits.

Of course, you already know this feeling. It's the reason we remind mothers in childbirth to let go of the contraction that has just passed. Take one day at a time, we whisper to those suffering heartache. This too shall pass.

When I am happy, possibilities are endless. I walk taller. Smile more. I lean over the fence to catch the attention of my neighbor. I plant things. Tidy up. Talk to small children on their bikes. Sing. Plan. Dream. Act.

And in that other place, the hopeless place, I feel my face contort with grief. I cannot bear the sound of joyful voices, any voices. I am a zombie. Stuck. Stiff. Fatalistic. When someone in a cherished relationship puts the relationship in danger, I feel the anguish of powerlessness. Something I love is being destroyed and I can do nothing to change it. I feel the same intense distress about the loss of other species and the destruction of glorious ecosystems. The chronic sorrow I carry in my soul flares and simmers, a constant heavyhearted companion.

EVERYTHING ABOUT CLIMATE JUSTICE is emotional. Extinctions of animals and plants, plastic pollution, the destruction of wild spaces, super-marine heat waves—each reflects and intensifies inequities between humans, and between ourselves and other species. Grief is the high cost of love that is lost or threatened. The obscenity of industrial animal agriculture, biodiversity loss and other forms of harm hits repeatedly in shocks of despair.

For me, it is the painful wrenching between the joy of loving the world and the grief of witnessing its destruction that makes climate emotions so difficult to navigate. I suspect the same might be true for you.

Over the past few years I have had the great fortune of collaborating with my friend Panu Pihkala, a gifted climate emotions researcher at the University of Helsinki. "It is not by accident that you specialize in hope," he tells me, referring to Robert Romanyshyn's book *The Wounded Researcher*. The research chooses us. We are drawn to study the things we need to work through.

Not long ago, I would have told you that my life has been shaped by this coexistence of hope and chronic sorrow. I might have shared the many ways I sought to intellectually understand and then learn my way out of emotions I did not wish to have. Yet in the last few years that has changed. I have come to value my emotions for the gifts that they are.

Reclaiming your emotions as powerful

"Happiness rewards us, sadness punishes us, and fear and anger elicit stress," writes James Dennison, an expert in political attitudes, psychology and behavior. I think this framing is too simplistic. The more I honor all of my emotions, the better able I am to appreciate the nuances and interplay between them.

For me, emotions are a source of wisdom. My emotions help me to feel when I am wrong or holding beliefs that contradict themselves. The discomfort forces me to accept it might be time for me to question my "truths" and look again. Joy, contentment and other positive emotions broaden my attention in a holistic way. They support an expansive, curious, creative feeling which fuels my desire and capacity to learn.

The breadth and range of emotions I am lucky enough to feel helps me to recognize and relate to how I am changing and growing, and what I value or am no longer willing to accept, just as the people, ideas, issues and places I care about change and grow around me.

The more my capacity to accept whatever I am feeling increases, the more reliably I can settle into a quiet sense of being grounded. It's a feeling of being able to count on myself, no matter what is happening around me. It's the opposite to how I feel when I'm emotionally upregulated and a surge of emotions takes hold. In those cases, I feel hurled to and fro by the unpredictable, confusing or painful behavior of others. In my grounded self, I feel braver and calmer to tackle whatever I'm facing rather than getting caught up in what I might wish was happening.

Positioning emotions as the opposite of reason is an outdated belief. A revolution in the science of emotions reveals how integrated emotions are with cognitive or logical thought. Emotions, it turns out, are the dominant driver of most meaningful decisions in life. Anger, for example, gives you the strength to respond to injustice, while anticipation of regret helps you to avoid excessive risk-taking.[1]

Climate change is emotionally charged

Appreciating that I am emotional—and accepting that talking about or acting on behalf of climate change generates emotional reactions in me and you and pretty much everyone else—is vitally important. Too often, we collectively act as if this isn't true. We teach or learn about climate change as if climate change is primarily a matter of scientific charts of greenhouse gas emissions or technological fixes. We operate in our daily lives as if we have a choice about what priority climate change should take and whether we might or might not choose to engage. We either think or don't think about climate change.

But all the while, the feelings we have about the state of the planet and what that means for our future or the future of our kids or other kids or species we love are screaming from the inside out. These stifled emotions show up as widespread eco-anxiety, cynicism, polarization, doomism and a host of other emotions.

Engaging with climate justice and other environmental issues is a high-risk activity from an emotional perspective. If you work in an environmental profession, or if you're a student or teacher of environmental subjects or a climate justice activist, you have a heightened risk for burnout. You are constantly immersing yourself in issues that you care about while navigating very real crises, along with a doomist narrative hammering home how broken things are. The academic research you read on environmental issues is also overwhelmingly focused on problem identification rather than solution generation.

The rise of climate emotions

In the past decade, a new academic field centered around the study of climate emotions has emerged. It brings together researchers and practitioners in varied fields of psychology, sociology, anthropology, public health, journalism and social science more broadly who are all looking at the emotional implications of the way we feel about the state of the world. It's a vitally important area, not only from the perspective of mental health but also for climate change action. As Susi Moser writes in *All We Can Save*, "Burnt-out people aren't equipped to serve a burning planet."[2]

A personal sense of empowerment is a fundamental aspect of psychological well-being and resilience. When you believe that you have some control over your life and that you can influence outcomes and overcome obstacles, you have a richer overall sense of life satisfaction. You feel more able to trust yourself to navigate challenges with confidence and purpose. This sense of personal agency enables you to make meaningful choices and create positive change. When you feel empowered, you're much more likely to engage with climate justice issues. When you feel defeated, you're unlikely to act, even on an issue you care a lot about.

Climate change is a social change issue. For me, a key question is: How do we remain in a stance of hope so we can empower ourselves to keep looking for and building upon proven solutions as we transform the destructive, inequitable systems that currently bind us? How do we hold hope to intentionally move toward just systems in which we and the 8.7 million other species on this Earth can thrive?

This pursuit of social change requires an unwavering acknowledgment of ourselves as emotional beings. It requires honoring how you feel as a means of identifying the very narratives that may hold you stuck. By including this stance, I wish to highlight three essential points for reflection:

Firstly, you are emotional. Your emotions have value and should not be dismissed.

Secondly, how you feel is how you feel. Your feelings are unique to you, *and* they reflect the narratives and cultures in which you are embedded.

Thirdly, you have agency with respect to how you perceive and navigate your emotions. Your emotions are powerful mobilizers of change and transformation.

Feeling your emotions in ways that support transformation

Panu Pihkala, the Finnish climate emotions researcher I introduced earlier, is a deep and beautiful thinker. He is quick to point out that you should not pathologize yourself or assume there is something wrong with you if you are experiencing eco-anxiety or other climate emotions. "Eco-anxiety," as Panu writes, is defined as "a chronic fear of environmental doom" and "the generalized sense that the ecological foundations of existence are in the process of collapse."[3] Anger, despair, fear, shame, hope, guilt—whatever feelings you might be experiencing are understandable and perhaps even healthy responses to the climate crisis. Your feelings are a response to your love of the planet, and the burden of what you are witnessing.

Naming and honoring climate emotions is a self-compassionate practice. Panu's work has inspired a Climate

Emotions Wheel you can find online (see resource list on page 273). It delineates a range of emotions that fall within broader categories of sadness, fear, anger and positivity. I use it to help myself and people I'm collaborating with to better understand and share our feelings. If you get a chance to take a look at it, ask yourself: What constellation of feelings am I experiencing? Where do these feelings reside in my body? Try to welcome everything you feel. Calm yourself with the assurance that there are no good or bad climate emotions.

Hozier's song "Nina Cried Power" is an exquisite expression of becoming aware of a problem, and the call to act on that awareness. It never fails to awaken the feeling deep inside me that knowing is not enough. It's what I *do* with what I know that matters.

Many people experience eco-anxiety as a series of awakenings. Often the stories begin with the first realization of the magnitude of the climate crisis or the imminent extinction of a beloved species. Many people I talk with spell out a lifelong chronology of eco-anxious feelings and thoughts that are currently influencing profound decisions, like the question of whether or not to have children in such a broken world.

You may have a story of feelings you want to tell. Other times, you may be unable to put your feelings into words. Revealing your innermost feelings, even to yourself, can be really difficult.

Anne Ylvisaker is an artist and writer.[4] She kindly shared the following "Words Feelings Art" practice for working with stories that you might want to express but are not yet able to voice.

Express your eco-emotions through Words Feelings Art

Gather colorful fine-tip markers, a small paintbrush and whatever color of paint you might have or could borrow. (You'll only need a little bit.) A piece of sidewalk chalk also works well. Cover a table (or whatever work surface you have) with newspaper or something else you don't mind getting paint or chalk on. Find a large sheet of paper. (Feel free to use the back of an old wall calendar or poster advertising a past event.) Place the paper in the center of the protected work surface. Make sure there is at least two extended hands' worth of covered surface beyond both the left and right edges of the paper.

Choose a comfortable place to reflect. Can you remember a time when you *didn't* know there was a climate crisis or species extinctions or plastic pollution or any other urgent, global environmental issue? What was your first awareness that something was terribly wrong? When did you start worrying about your future or the future of your kids, or feel bewildered or overcome by senseless destruction?

When and if you feel able, find a word or short two- or three-word phrase that captures the emotions that are most upon you at this moment.

Using your paintbrush, write that word or phrase in a large script beginning on the work surface, continuing across the paper, and finishing on the other side of the work surface.

Now lift your paper off the work surface. You will have part of a word or phrase, full of that emotion that only you will know is there.

Look at the shapes made by the partial letters on the page. How might you add words or drawings to those shapes to express the emotions of your story? Choose whichever markers are calling to you, and let yourself express those feelings through lines, colors, words, more paint—it's your choice entirely.

Let this artwork emerge as an intimate expression of your eco-emotions just for you.

New words to express climate trauma

Be mindful if you are experiencing paralyzing grief, trauma, depression or other mental health concerns, or if someone you care about is experiencing these things. Some scholars use the term "pre-traumatic stress" to describe the worry, fear or uncertainty you may feel about climate impacts already happening combined with those predicted in the future. Your eco-anxiety may be heightened or triggered if you directly experience heat waves, floods, wildfires or other climate events. The approval of a destructive project or a dire headline about the failure of a major climate change negotiation can spark secondary or vicarious trauma. New theories about "climate trauma" and "ecological trauma" give us language to face and talk about our grief and other feelings and experiences, and to seek the support we need.

Your brain is not a
computer hardwired for fear

How often have you heard the analogy that your brain is a computer and your culture—your values, know-how and customs—is the software? This widespread idea has long passed its expiry date. Your brain is not a computer. You are not hardwired for fear, or anything else, despite how frequently this so-called fact is repeated. Writing in a 2023 journal article, neuroscientists Cassiano Ricardo Alves Faria Diniz and Ana Paula Crestani make it clear that a "hard-wired" conception of the brain is simply outdated science:

> For so long the adult brain was considered hard-wired, incapable of any accommodation. However, after some breakthroughs, neuroplasticity is now well recognized as a fundamental and lifelong brain property.[5]

Neuroplasticity is the ability of your brain to form new connections and pathways in response to new circumstances. Not only do your existing cells have the capacity to interrelate in new ways, your brain can also grow whole new neurons—a process known as neurogenesis.

Indeed, a lot of what you've been told about the brain is riddled with myths. When you face a threatening situation, your lizard brain automatically activates the fight-or-flight reactions, right? Wrong. You do not have a lizard brain—a small, primitive part within your brain that functions entirely on instinct. "The only animal on this planet with a lizard brain is a lizard," says Lisa Feldman Barrett, professor of psychology at Northeastern University. "Scientists have known since at least the 1970s that the idea of a lizard brain is a fiction of neuroscience."[6]

Before I go further, I want to point out that this "lizard brain" myth not only distorts the reality of what is happening in your brain. It also misrepresents the capacities of lizards. It reinforces an equally out-of-date myth that these animals are cold, stupid and unfeeling—prisoners of predetermined instinctive responses. On the contrary, lizards—and reptiles more generally—are highly emotional, often social animals with complex forms of parenting, courtship and nesting. Study after study reveals what the "lizard brain" label prevents you from seeing: Reptiles are a lot smarter than you might have ever thought.[7]

The reason these are such important myths to dispel from the standpoint of hope is that they rob you of agency you actually have but of which you may not be aware. Your emotions are beautiful parts of you that help you to experience moments of delight and lifelong loves. They help you recognize what matters most to you, or when the circumstances you are in just simply aren't right. Your exquisite brain is constantly creating your emotions in response to physical sensations in your body, the surroundings you're in, and your past experiences. The making of your emotions is an ongoing dance, not a fixed, stuck state.

This dynamic capacity is vitally important to navigating climate justice and other urgent, global issues. According to an international team of neuroscientists and psychologists writing in a 2021 issue of *Neuroscience and Biobehavioral Reviews*:

> In the present global context of active threats of climate change, pandemics, and growing economic disparities and inequities, mental health concerns are rising worldwide. Considering the backdrop of these imminent

societal challenges, advancing scientific research that focuses on wellbeing and healthy emotional outcomes is crucial. The experience of positive emotions, feelings, and affect are fundamental building blocks for cultivating resilience, flourishing, vitality, happiness, and life satisfaction, which ultimately contribute to physical and emotional wellbeing.[8]

I find this deeply empowering. The emotions that seem to happen to you are actively made by you and the circumstances in which you are living.

I am a passionate person and I often struggle to rebalance myself when my emotions are running high. I am drawn to the concept of equanimity, which I understand to be the capacity to see with patience and understanding so that I don't take offense at things when they aren't personal, and so I can choose how I might wish to react. Cultivating equanimity better enables me to calm and center myself no matter what life brings. I find it helpful to keep practicing ways to find a calm center in day-to-day situations that are frustrating, but pretty low stakes. That is how I chanced into the following practice.

Cultivate equanimity, or "Speak kindly to your dog when they are barking"

I have a chihuahua, which means I engage in this practice several times a day! Even after fourteen years of living together, when Cookie Dough (that's her name) starts barking, I can still find myself tempted to start hollering at her to be quiet. Isn't that nonsensical? To yell at someone to be quiet? And yet at times I find the barking so annoying, it's easy to lash out. The practice is making the decision not to yell, but instead to kindly ask her to make a better choice.

Speaking kindly to your dog is a much more respectful way to treat another living being. Plus, adding my loud voice to the barking simply increases the annoyance for my neighbors. Yet the reason I find this practice so valuable is that consciously pausing and deciding to speak kindly to my dog when she is barking reminds me that pausing and deciding is the best way to handle any situation in which I feel emotionally charged. Hearing myself speaking kindly also shifts my own feelings. I end up feeling calmer and more confident that I can count on myself to do the right thing even when it is hard. Speaking kindly to my dog when she is barking helps me shift an ongoing frustration into the fuel of empowerment.

Whether or not you have a barking dog, I'm guessing there is something that happens in your everyday life that you find emotionally charged. Choose to meet that with a verbal expression of kindness. Watch to see how difficult this is to do. Feel how it positively affects your sense of self when you are successful.

Diversifying eco-emotions
beyond what is WEIRD

Your emotions are unique to you and to the cultures with which you identify. To me, that's both an obvious *and* mind-blowing thought. I know we are all individuals, and yet, I think I've been operating upon the assumption that how we feel as humans is universal.

You would be forgiven if you also thought the same way. After all, almost all the psychological research you've been privy to has operated on a similar premise. It turns out to have been based on a very specific population of people and then extrapolated to apply to all humans on Earth.

The "standard subjects" involved in these studies are anything but standard. They are predominantly U.S. college students who sign up as research subjects in exchange for money or course credits. In June 2010, human evolutionary biologist Joe Henrich and his coauthors published a landmark paper revealing this flaw. "Behavioral scientists routinely publish broad claims about human psychology and behavior in the world's top journals based on samples drawn entirely from Western, Educated, Industrialized, Rich, and Democratic (WEIRD) societies," they wrote.[9]

More problematic still, WEIRD people are outliers. Because success in cultures that center the individual, like the United States, requires each person to stand out, the researchers argue that WEIRD people are more focused on selling themselves, and thus, they pay more attention to personal achievement, individual attributes and dispositional virtues than to people's roles and relationships in the wider community, or to the situations they are in.

I had a personal experience back in 2000 of what happens when data from one group of people is generalized to a different population. I was pregnant with my son and I was interested in running marathons throughout the pregnancy. My obstetrician's advice was not to exceed a running heart rate of 140 beats per minute. Yet no matter how slowly I ran, I couldn't keep my heartbeats down to that level. I worried constantly that my running might be harming my baby. That heartbeat recommendation has long since been removed by the American College of Obstetricians and Gynecologists, yet it appears to still be commonly shared as folk wisdom. It's such a great reminder of why it's essential to double check for outdated content. My worries were completely unnecessary. Years later I learned that female subjects were so significantly underrepresented within sport and exercise science research that the heart rate advice for pregnant women I received was extrapolated from research on, you guessed it, WEIRD men.

Only 12 percent of the world's population is WEIRD. Yet a staggering 95 percent of samples among top psychological journal articles from 2014 to 2017 were drawn from the WEIRD population. Less than 1 percent of psychological studies included African research participants, despite Africa being home to almost one-fifth of the world's population. A follow-up study in 2020 revealed that the needle on this issue had barely moved.[10]

This revelation is painful and, sadly, might not come as a shock. It's yet another example of how the long shadow of colonialism continues to skew our perception of how people around the world perceive, think, feel, make decisions and relate to each other.

What I find hopeful about this blatant example is that it illustrates how often "deeply ingrained truths" turn out to be based on erroneous assumptions. Once you expose the assumptions, you are no longer bound by a particular false idea about the way the world works. More inclusive approaches that recognize the incredible complexity of life on Earth open up. Change becomes more possible as faulty assumptions that hold old "truths" in place are exposed.

What is emerging is a more diverse picture of human nature. A much larger proportion of cultures around the world, for example, turn out to be more focused on relationships—friendships, allegiances and family alliances—than on the individualism personified by the WEIRD participants. Ellen Morgan, a scholar in the Department of Social Intervention and Policy Evaluation at the University of Oxford, writes that people "from so-called collectivistic or relational cultures—such as those often found in Asia—have more interdependent views of themselves... [they] tend to see other people and their relationships with those other people as a part of who they are."[11]

Culture drives
how you think and feel

We humans are fundamentally a cultural species, which means that different cultural values create different emotions and attitudes.[12] Every society has a dominant set of norms that influence which emotions are seen as appropriate, how we should regulate our emotions, and how emotions are entitled to be expressed. How you learn as you navigate the cultures in which you live actively changes your brain, your hormones and other parts of your biology.

In other words, your culture shapes not only *what* you think but *how* you think and perceive the world.

Climate justice and other environmental issues kindle emotions that also vary depending on what matters to you personally, your circumstances, the weight of multiple injustices you may be experiencing in your life, and many other factors. By refusing to pathologize or label individuals who are experiencing eco-anxiety, you keep the attention where it must be: on the broader, structural issues that cause and disproportionately distribute the terrible impacts of climate change.

Just as the negative impacts of ecological crises do not affect people equally, the ways in which each person experiences eco-anxiety and other climate emotions depend on gender, geography, marginalization, culture, socioeconomic status and more.[13] Eco-anxiety is a global phenomenon that cuts across privilege, race and geography, *and*, as my friend and environmental humanist Sarah Jaquette Ray writes: "It makes sense that people experiencing climate change the most would feel the most anxiety about it. It also makes sense that youths around the world are feeling the most dread about how climate change will unfold, especially in their lifetimes."[14]

Writing in the journal *Frontiers in Psychology* in 2024, researchers Shicun Qiu and Jiacun Qiu challenge the tendency to focus too narrowly on the individual experience of ecological emotions. Drawing insights from "the holistic and collective wisdom of indigenous cultures and Chinese traditional philosophy," they demonstrate that when we move beyond the limitations of Western individualism and experience our profound connections to each other and other species, we are better able to see the potential of our ecological emotions to foster a collective consciousness—a motivation for community engagement.[15]

We see this in the way so many people love and respect old-growth forests. It's as if we are part of a massive emotional community that collectively feels fiercely committed to their well-being. Those emotions cause us to rise up together. Americans submitted more than a million comments between 2022 and 2024 demanding that the United States Forest Service protect old-growth forests.[16] "In 2021, Fairy Creek became the scene of the largest civil disobedience action in Canadian history," writes Shannon Waters in *The Narwhal*. "Following more than 1,100 arrests, and at the request of Pacheedaht First Nation, the B.C. government deferred just over 1,180 hectares of Fairy Creek old-growth forest from logging."[17]

Paying attention to your individual feelings while at the same time recognizing how they are influenced by the communities and systems in which you live is an empowering combination. I often feel powerless when I experience inequality. Yet when I acknowledge these feelings and then shift my attention to the systems that create the injustice I am experiencing, I am sometimes more able to envision possible avenues for change. That's because inequitable systems, like all systems, are human constructs. They may be so deeply embedded they feel impossible to change, but they are nonetheless alterable.

Panu Pihkala has created a model of how people experience eco-anxiety and ecological grief that is very helpful for understanding the dynamic processes of accepting, adapting to and coping with living in the climate crisis.[18] Over the past few years Panu and I have worked together on a number of projects, most recently in an intergenerational collaboration with climate influencer Queer Brown Vegan, to translate Panu's academic model into formats that are easily shareable through social media. For instance, as I write this, I've just finished co-creating a short animated film based on the model called *The Dance: Living With Eco-Anxiety*, which is freely available online.[19] The film was directed by Spotted Fawn Productions, an Indigenous-led animation company founded by Emmy-nominated filmmaker Amanda Strong, who is Michif/Red River Métis. The soundtrack for the film was created by PIQSIQ (pronounced "pilk-silk"), sisters Inuksuk Mackay and Tiffany Ayalik, who are Inuit-style throat singers from Arctic Canada.

Through this creative intergenerational collaboration we have been growing our understanding—and the representation of eco-emotions—beyond the white and Western demographic roots of academic research. We've also been expanding our thinking to collaborate with other species. When I was writing the script for *The Dance,* for example, I took myself out for walks in the redwood forests near Santa Cruz, California, listening for inspiration among those very old trees. One day, a coyote darted across the path, so close I nearly tripped over her. "A trickster!" I thought. At that moment, I had the starting place for the film: "Climate change is always with us. Fear is a trickster, hungry to grow." Amanda translated that spirit into illustration as a felt presence that permeates the film.

Joining together in collaborations exposes the taken-for-granted norms that govern each of our individual lives. If, like me, you live in a predominantly individualistic, human-centric society, you may automatically respond to eco-anxiety or other challenging emotions by trying to boost your own resilience. You've been conditioned to do this by a culture that pushes climate action onto individuals, and holds you accountable for your lifestyle choices.

Seeking answers within yourself is valuable (as the practice on page 112 to grow your capacity to shift your emotions in more empowering directions demonstrates). And, at the same time, questioning this impulse to focus inward also has great merit. Working with people who live in more collective societies reminds me that individualism is not a universal norm. No matter how hard you try, you can't solve a global issue that affects all humanity on your own. The more often you expect yourself to do so, the more likely feelings of anxiety, failure and despair will grow. Be aware of your own actions *and* name the systems that enhance or restrict your ability to behave in the ways you know are right. Reach out for support from other people, and from the greater ecosystems of which you are an intimate part.

Emotional self-care and collective care

What helps you to continue when you feel overwhelmed or hopeless? For me, it's kind words and genuine offers of care. "I will help you." "I've got your back." These simple phrases offer tremendous power to comfort me. Compassion and care show up time and again in the research literature as vital for mental health and well-being.

Bolstering relationships with others who may be experiencing many of the same emotions is healthy. Climate cafés where people come together to share feelings and act on climate change are emerging in cities all over the world, with lots of virtual versions on offer as well. I appreciate the mix of grounding mindfulness practices and camaraderie that I have experienced when people gather to share their fears, anxieties and other emotions about the climate crisis.

Reminding yourself of support you can count on can really help you in dark times. The following practice of self-care, called Circles of Support, comes from Jennifer Atkinson, professor of environmental humanities at the University of Washington Bothell. Jennifer adapted it from work done by Bob Doppelt in his book *Transformational Resilience*.

I met Jennifer just before the COVID-19 lockdowns hit. We were co-leading a project with Sarah Jaquette Ray, chair of the Department of Environmental Studies at Cal Poly Humboldt, to bring together scholars and practitioners interested in climate emotions. We had planned a small gathering of people who were to be kindly hosted by the Rachel Carson Center for Environment and Society in Munich, Germany. When the lockdowns prevented us from traveling, we made what turned out to be a very fortuitous decision: We created an online event and opened it up to the hundreds of people who had initially expressed interest. The international network that emerged remains active and has co-created a book, *The Existential Toolkit for Climate Justice Educators*, which you can freely source online. You can learn more about Circles of Support, and its origins, among the online resources (see page 273).

Gather strength and resilience
from your Circles of Support

Take a blank piece of paper and write your name in the
center.

Draw a circle around your name. In that circle, draw or
write down your personal strengths and skills that you can
rely on to help you deal with adversity.

Draw a second circle around the first. Within this second
circle, draw/write the internal mental resources that make
you feel safe and calm. Ask yourself, *What brings me internal tranquility?* (It might be images of the ocean, or seeing
yourself walking in a forest, or spending time in a spiritual
or religious practice.)

Draw a third circle around the second. In this one, draw/
write the names of your true allies—friends and family
(don't forget favorite dogs or cats or other species) whom
you can depend on no matter what.

Now add a fourth circle. Draw/write the special places
and external resources that help you to feel safe and calm.
(Perhaps your friend's couch, money in the bank, a quiet
garden...)

Finally, encircle these beautiful, very personal circles of
support with a fifth circle. Draw/write the ecological systems, organisms, or processes that make all of this possible.
This is my personal favorite circle because it is the one in
which I can express the never-ending support I can count
on from the sun, the forests and the ocean; the magnificent process of photosynthesis that creates the air that I

breathe; the birds that spread the seeds that create the food I eat...)

Keep your Circles of Support near to hand. Accepting not only the truth that you are living in a climate justice crisis but also the difficult emotions that come with that, and finding meaningful ways to cope, is an ongoing quest. Draw upon your inherent strengths, as well as the important people, other animals and plants, and processes you can trust for support, as often as you need

Transforming your emotions to empower change

To nurture hope, it's important to recognize the multitude of factors that influence how we feel. This complicates the erroneous perception that how we feel is a direct response to taken-for-granted truths. Reflecting on our own emotions opens new avenues for us to reengage with the issues that we care about but might have categorized as already doomed.

You have agency with respect to your emotions. As you reframe your ideas and expectations, you may choose to foreground certain emotions and set aside, suppress or alter others in service of the changes you seek to make. Choosing to be hopeful, for example, requires you to acknowledge whatever feelings of fear or worry or other emotions you might also be experiencing, and to hold them safely while you lead with hope. At the same time, you may need to recognize and challenge the ways in which cultures of fear are created and used to sustain the status quo. Choosing to be hopeful, therefore, is a decision to engage in deep transformations that are emotionally challenging because they involve overarching shifts in social, political, cultural and personal aspects of your life.

Social and political arrangements shape our emotional lives and, in turn, our emotional experiences and practices can entrench unjust social structures, or resist and transform them. Just as critical race scholars outline ways to transform guilt into active anti-racism, it is important to consider ways to transform emotions rooted in climate privilege (unearned circumstances that insulate someone from the worst effects of climate change) into climate justice action. Writing in the *New York Times*, Carola Rackete, an ecologist and social justice activist, says:

> Many, particularly those of us in the white middle class, are not used to fighting uphill battles against unequal power structures. Many of us have not been taught how to build community and collective power in a situation where the odds are stacked against us... But I know that my privilege gives me responsibilities not only to communities struggling for their survival, but also to the global community of all living beings. The fight for global climate safety is now at our doorstep. To succeed, it will need a culture of resistance and a clear vision of justice and solidarity.[20]

Your emotions motivate you to act. They fortify you through the long, hard work of changing things for the better, whether on a personal or collective level. "I am emotional" is an essential stance because emotions drive and sustain your capacity to make the changes that matter most to you.

Decide, why, when and how
you want to regulate your emotions

How you choose to express your emotions is a deeply important question that deserves thoughtful consideration. Given the essential role of emotions, is there a danger in consciously deciding to be hopeful when you are also feeling worried? Is it a problem to try to shift how you feel?

If regulating your emotions is imposed upon you by force or other forms of emotional abuse, it is clearly wrong. If you have been regulating your emotions because of social pressures of which you might previously have been unaware, then give yourself time and space to consider what influence these pressures have on you. Regulating your emotions has to be a choice you make freely.

Large-scale social change is fraught with many different emotions. It may involve radical shifts in various aspects of your life and your understanding of how you and the rest of the world work. In every moment, you are simultaneously remembering change, adapting to change and imagining change. At the same time, your emotions are influenced and shaped by the stories or "narratives" you tell yourself, and the narratives you are told by the broader societies in which you live. That is why I am calling out the power of fearmongering and doomism, and why I am positioning hope as a political choice. You may hold onto hope because you feel the relief of seeing a species recover, or you may hold onto hope because you resent the systems of doomism that seek to manipulate you into disempowerment. Emotions related to transformation are as varied as the experiences of change.

You may encounter some aspects of social change with excitement and enthusiasm, and others with sadness, powerlessness and anger. You may feel lost, insecure, disappointed and anxious. All of these feelings are understandable and valuable. The way you feel is the way you feel. Your emotions are deeply ingrained in your body and in the ways you relate to others and the world around you. Adjusting to social change involves self-transformation. That is why this chapter involves practices that relate to how you think, embody, care for and express your inner feelings.

Your brain creates your emotions on demand

Lisa Feldman Barrett's research and writing has had a big impact on me. "Emotions are your brain's best guess of how you should feel in the moment," she writes. "Emotions aren't wired into your brain like little circuits; they're made on demand. As a result, you have more control over your emotions than you might think."[21] I appreciate how clearly Lisa helps me to question the dated view that emotions are triggered by stimuli and beyond my control. Instead, she helps me to see that they are generated by my brain and the rest of my body in response to whatever situation I am in. I can honor and understand where they come from while at the same time not taking them to be inevitable or unchangeable.

Like most people I know, I feel triggered when a painful experience reminds me of painful experiences I've had before. What I've come to realize over many years of practice is that even though I still experience the same overwhelming explosion of adrenaline surging through my body, I am sometimes a wee bit better now at choosing to do something different with these very upset feelings. At first, the best I could do was make myself go outside. In fact, at the very beginning, the best I could do was make myself look toward the door. Yet with a lot of practice and many failed attempts, I'm more able to see my anger, for example, as a sign that something I care about is being treated unfairly. I am learning to appreciate my anger. My anger is my voice of justice.

Knowing this motivates me to try to shift from anxiety to determination, or from fury to power. I'm consciously trying to develop my capacity to decide how I can speak up for the important thing that is under threat in the midst of my anger, rather than getting swept along in an outburst that just leaves me depleted. Over time, my brain is learning to respond in ways that give me more agency and a wider repertoire of possible actions. Developing this flexibility is key to resilience in any situation, even the most stressful ones.[22] When I manage to carry it off, it is a relief to me to feel energized rather than squashed by my emotional response to something that matters to me being at risk.

I use the practice on the following pages to help grow my capacity to shift in emotionally charged situations toward more positive emotions that support my resilience, well-being and empowerment. I've used anger as the emotion in this particular example, but you can use the practice for any emotion. Feel free to tweak the practice in any way that best works for you.

Shift toward more empowering emotions— a practice in five parts

Part one: Focus on your body

The moment you start to feel upset, run a check on your body. Ask yourself:

- **What is the emotion I'm feeling?** (Try your best to name it.)

- **Where am I feeling it in my body?** (Nausea in my stomach? Constriction in my throat? Clenched hands? Shallow breaths? Shaky legs? Tight jaw?)

- **What energy do I feel?** (Numb? Like throwing something? Wound up? Exhausted?)

Stay with the emotion and "listen" to what it's trying to share with you without overthinking or judging:

- **What message is my body sharing with me about what it needs?**

- **What connection is it seeking?** (A hug? A break? Validation?)

- **What do I physically need at this moment?** (Some food? A warm blanket? To run? Fresh air? To be quiet?)

The physical sensations in your body influence what your brain guesses you're going to need in the next moment, which influences your emotional response. Identifying and naming what's happening in your body can help you to choose the next best step to take.

Part two: Focus on your surroundings

If you continue to feel overwhelmed, try to notice anything that you are hearing, seeing, smelling, tasting or touching that is amplifying your feelings. Can you change any of these things? (I've discovered that bad lighting and loud construction noise increase my irritation.)

Take yourself outside. Give yourself a "time out." Move rocks. Weed a garden. Change your current situation in any way that enables you to release unwanted energy. These changes have the potential to influence your brain's unconscious guesses and thus alter your emotion.

Part three: Consider how your past experiences shape your emotions

No matter what you are experiencing, your brain will try to create an appropriate emotion, likely the same emotion as the last time you were in a similar situation. This is what makes shifting your emotional responses so difficult. If you've been hurt in the past and you're experiencing the same feeling again, trying to regulate your emotions can feel like exactly the *wrong* thing to do. If anything, you're probably increasing the volume to make sure you are being heard. You want the hurtful behavior to STOP.

There are many ways to regulate your emotions. I am attached to one that, sadly, doesn't work. When I'm angry, I tend to curate examples of all the times I have felt these feelings in the past. Psychologists call this "rumination" or "carpetbagging." By concentrating on negative situations, I find the intensity and length of my anger actually *increases.* An hour later, I'm angrier than I was when the incident occurred. I understand rationally that it is a deeply flawed approach. But in the moment, I feel as if I am providing evidence for why the situation is wrong and how desperately it needs to change. I am especially prone to do this when the person I am arguing with is using "suppression" to regulate their emotions.[23] The more they downplay, avoid or refuse to acknowledge the problem, the more I layer on the negative examples.

If this is happening, accept whatever emotion you are feeling. Know that it comes from an authentic place. Try to remember that when you are in the midst of an unregulated emotion, you can't think as clearly. In this highly charged moment, each thought of something or somebody that did you wrong kicks off another physiological reaction. That's why the practice of regulating emotions needs to begin *before* you are highly charged and continue *after* you have found your calm.

Part four: Cultivate new experiences

Past experiences are the most difficult part of the practice to address because you can't change what has already happened. But you can change your present, which puts your brain in the position to guess differently in the future. When you are in a calmer state, make a short list of questions to ask yourself to reappraise your emotional responses. List ways to cultivate new experiences that differ from the way you responded in the past.

I love to listen to podcasts in which people talk about their lives. They fuel my appreciation for the varied ways people across cultures and circumstances experience emotion and navigate injustices. When I hear a new way of experiencing sadness, for example, I actively try to embody it. In a BBC radio interview, Zarifa Ghafari, who was Afghanistan's youngest female mayor when she was appointed in 2018, described the heartache of her father's assassination, followed soon after by the loss of her country.[24] From her, I learned about the capacity for generosity in the midst of sadness. "There are two Zarifas in one me," she said, explaining how she would cry through the night and then come back to her family home in the morning with a smile to give confidence to her mother and siblings. This "trying on of other people's emotional responses" has become a way to feed my brain new experiences that shape my own emotional responses.

Part five: Create moments of release
Your emotions are complicated and rich with wisdom. You feel them rapidly and they can drive swift action. Feelings are meant to share their message and leave your body. The goal is to benefit from what they are telling you without getting triggered into a loop of reactivity. That's why breathing exercises and other embodied practices are so helpful. For me, a walk on a blustery day is an ideal way to release anger. Pushing against the wind as branches career above allows me to express and then liberate my rage.

Take a moment to jot down the most effective ways you have found to release your own challenging emotions. "Some people keep a written journal of things they're grateful for every day. Other people meditate, which teaches them to be compassionate," writes Lisa Feldman Barrett. "The more often you cultivate new experiences, the more you build up your brain's toolbox for making emotions in the future."[25]

HOLD YOUR HOPEFUL STANCE while still remaining open to experiencing every other emotion you might have. Hope is a way of choosing to be in the world. Your feelings are a reflection of how you feel in the world at any given moment. Let them all coexist. Welcome your frustration, your sorrow, your anger while you hold your hope, all at the same time. You can choose. And the reason you can choose is that you have a lot more influence over how you feel, and in turn, what would help you to act on behalf of climate justice, than you might imagine.

STANCE 4

I Am on a Reparative Quest of Transformation

WE SAT ACROSS from each other, shouting to introduce ourselves above the happy clamor of a busy Friday night pub crowd in Wimbledon. I think he told me about his trip from the airport, but I couldn't quite make out the details. It was too noisy, plus my attention was diverted by a whippet who magically appeared on the bench beside me, smuggled in beneath the puffy jacket of his owner.

It was the next day, when he took his place at the podium, his friendly, gentle voice filling the room of the conference we were both attending on Challenging Hope, that I began to register how easy it is to miss the presence of an extraordinary person. Even when you're sitting knee to knee in a bar.

Munyurangabo Benda is a tutor with the Centre for Black Theology at the Queen's Foundation in Birmingham. He hadn't started out to be that—his ambition was to be a lawyer, most likely a diplomat, eager to join the politics of his country as it emerged from civil war. That dream was not to be. His country was Rwanda, and before he could complete his law degree, one of the most notorious modern genocides broke out. Over one hundred days between April and July of 1994, nearly one million ethnic Tutsis and moderate Hutus were killed as the international community stood by.

In steady, patient language, Benda (as he asked to be called) shared that experience with those of us gathered in the room. He was kind in his telling of the horrors, careful to

118

ensure our emotional safety. The story he told is the story of what happens when a culture of fear transforms into the politics of atrocity.

"Genocides happen when people refuse to commit. When they withhold their agency. Their generosity. Why do they withhold? Fear. The message is clear: Stay away or else. Most people do not commit because the people who are being targeted fall out of the magnetic field of their moral concern. Under terror most people will comply."

So often, people ask me if hope is a privilege. On that day, as I sat in silence, willing my compassion toward the man generously sharing the unfathomable so I and the others could try to understand, I knew what I have always believed. Hope is not a privilege.

"The problem isn't trying to have hope in the actual moment. Hope is easy. This thing called life truly wants to live. Through one hundred days of constant threat. As you see people dying, you believe this is the end of the world. You need hope to live in the aftermath."

I found myself looking at his hands. Soft brown hands. As he described the open slaughter, an intimate type of violence that was not in the scope of comprehension, I kept thinking about the tenderness of touch. Touch the hand of someone you love and your breath and heartbeats fall into the same rhythm.

Hands that soothe. Hands that butcher.

"Hope exists because, you see, I'm here. A lot of work has gone into me. We lived after the genocide because during the genocide there were stepping-stones of hope. Prayer and spiritual rituals. Incomprehensible acts of people who stood in the way of death at the risk of death."

"I lost my girlfriend to the genocide. I lost love as well."

"I found hope in mixed marriages between Hutus and Tutsis. Despite violent opposition these young people couldn't wait to get married."

"Why do I tell you this? What has been must be spoken about. Some violences need to be communicated. Lamented. There is no tradition of how humanity shares these atrocities."

AM HUMBLED by the capacity of people to continue to live beyond tragedy. Benda's generosity in sharing his story of personal trauma in the service of greater understanding fills me with reverence and appreciation. I am deeply grateful to him.

For many years I have listened to the stories of people bravely sharing their voices on the Forgiveness Project, a portal co-created by the late Archbishop Desmond Tutu. I have sat in circles, full of awe and sorrow, as my late friend Lillian Howard, a member of the Mowachaht/Muchalaht First Nation, found the courage to share her experiences as a residential school survivor and victim of domestic violence. Choosing to recount those traumas came at a personal cost, yet she did it in the service of voices that were silenced by the cultural genocide perpetuated by the government of Canada through the Indian Residential Schools.[1]

IN 2018, a thirty-year-old Southern Resident killer whale (*Orcinus orca*) carried her stillborn calf for seventeen days across hundreds of miles. Tahlequah, or J35, as she has been named, carried her dead baby on her head and her rostrum.

She was not alone in her grief. Resident killer whales live in close-knit family groups. Throughout this tragic period of loss she was accompanied by other pod members who took turns supporting her and the calf. Southern Residents, a distinct community in the Salish Sea of the Pacific Northwest, are highly endangered due to vessel traffic noise, environmental contaminants and overfishing of their main food, salmon.[2] "We've been yelling into the wind till the mother whale kept her baby up and the whole world saw," says Ray Harris, a member of the Stz'uminus First Nation on Vancouver Island. "The whales woke us up."

SOMETIMES HOPE IS A BURDEN. Hope is a push and pull. Pushed back by fear and despair. Pulled forward by justice. I believe there is much to be learned from these deeply moving examples of betrayal and grief and courage and hope that matters to the quest for climate justice. The metamorphosis toward climate justice demands relational repair on a personal and societal scale. That is why embodying this stance—*I am on a reparative quest of transformation*—is so important.

What I have learned (and have to keep relearning) is that all reparative quests begin with, and depend upon, truth telling. It is not possible to step beyond betrayal until the truth of wrongdoing is acknowledged. The person or institution who violated the trust must display a genuine commitment to repair what's broken. Whoever has been harmed may need space and time and support to express and process their emotions and to decide if, when and how they might move beyond this profound loss.

What is true in our personal experiences of betrayal is also true for anyone suffering from institutional betrayal and the physical and emotional wounds of climate change.

The tendency to treat climate justice as primarily a question of adopting effective technological fixes or enacting much-needed political will ignores the depth of suffering it is causing. Healing from the ravages of climate injustice demands what all healing requires: truth telling, acknowledgments of wrongdoing, apologies, making amends, disrupting the broken systems that hold injustices in place—and the consideration of forgiveness. Authentic actions of repair are necessary for someone to open their heart to the risk of believing in and engaging with climate justice transformations. Equity moves at the speed of trust.

Expressing the emotions of broken trust

If you are twenty-five years of age or younger, you have experienced betrayal with respect to climate justice throughout your entire life. I imagine you may already feel this deeply. According to the largest international survey of climate anxiety to date:

> Children and young people in countries around the world report climate anxiety and other distressing emotions and thoughts about climate change that impact their daily lives. This distress was associated with beliefs about inadequate governmental response and feelings of betrayal. A large proportion of children and young people around the world report emotional distress and a wide range of painful, complex emotions (sad, afraid, angry, powerless, helpless, guilty, ashamed, despair, hurt, grief, and depressed). Similarly, large numbers report experiencing some functional impact and have

pessimistic beliefs about the future (people have failed to care for the planet; the future is frightening; humanity is doomed; they won't have access to the same opportunities their parents had; things they value will be destroyed; security is threatened; and they are hesitant to have children).[3]

These heartbreaking findings mirror the results of a previous study, in which young adults were asked about their experiences of learning about climate change. They described feeling "stranded by the generational gap... frustrated by unequal power, betrayed and angry, disillusioned with authority, drawing battle lines."[4]

Feelings of broken trust are heightened by the failure of governments and other institutions, and by the actions of previous generations. As the influential *Youth Climate Justice Handbook* describes: "The climate crisis is an intergenerational injustice of existential proportions. Young people and forthcoming generations will be disproportionately affected by the devastating effects of climate change."[5]

What we are talking about is betrayal. And a failure of the institutions who are charged with caring for us and the planet to tell the truth, and to take responsibility for changing the unjust systems that do not serve the vast majority of people, other species or the planet itself. That is why so many are responding in anger.

Anger is a powerful response to wrongdoing

As I wrote earlier, anger is an activating and complicated emotion. Anger lies at the heart of great wrongs—from domestic violence to war. Anger, at least in part, also drives nearly every positive social movement, from disability rights to civil rights. Anger is an ethical response, an emotional response and an action response.[6]

I believe that anger is valuable. I believe that anger is necessary in the pursuit of justice. Given its power, I encourage you to think carefully about what you personally wish to do with your anger. How best can it be used to drive the justice transformation?

Lots of people have meaningful ideas about the role of anger that you might want to consider as you think about how you wish to show up in the world. "As a passionate activist, I'm often surrounded by those who feel extreme anger towards institutions and oppressors that continue to create and inflict severe pain," writes youth activist Leyona Bray Kaji, who self-identifies as a white-passing woman in higher education living in the U.K. "I, too, feel rage towards those responsible for denying basic rights whilst destroying the planet, but I do not believe spreading anger is the solution. I want to see widespread concern about these issues, individual behavioural change, and mass civic engagement. I don't want to see red. Not out of ignorance but because I'm unsure whether anger is actually effective."[7]

For Indian author Pankaj Mishra, anger is a well-founded response to the history of inequity that shapes our current global crises. In *Age of Anger* he writes, "The two ways in which humankind can self-destruct—civil war on a global scale, or destruction of the natural environment—are rapidly converging."[8]

Anger is a justified emotional response to wrongdoing, says Martha Nussbaum, professor of law and ethics at the University of Chicago. Yet on its own, it lacks the capacity to enable us to solve problems. Too often, its value as a response to injustice is hampered by a wish for payback or revenge. The belief that violence or inflicting proportional pain on the person who committed the harm will somehow balance out the offense is a fantasy, she says; the notion of payback is "deeply human but fatally flawed."[9] So too is the way in which anger can become focused on saving face. An obsession with personal honor impedes us from attending to the needs of the collective. If we actually want to do something good for ourselves and for others, Nussbaum argues, non-anger is far more useful.

While vengeful anger deepens harm, unselfish anger at injustice in the service of victims of wrongdoing can be a powerful force for change, especially when it is oriented toward changing laws, policies or institutions. "As heat conserved is transmuted into energy," wrote M. K. Gandhi, "even so our anger controlled can be transmuted into a power which can move the world."[10]

During his twenty-seven years in prison, Nelson Mandela meditated on the role of anger in societal transformation. In his book *Conversations With Myself* he shares a story about an argument between the sun and the wind, which he often used to advocate for nonviolent routes to peace:

> The sun said, "I'm stronger than you are" and the wind says, "No, I'm stronger than you are." And they decided, therefore, to test their strength with a traveller... who was wearing a blanket. And they agreed that the one who would succeed in getting the traveller to get rid of his blanket would be the stronger. So the wind started. It started *blowing* and the *harder* it blew, the *tighter* the traveller pulled the blanket around his body. And the

wind blew and blew but it could not get him to discard the blanket. And, as I said, the *harder* the wind blew, the *tighter* the visitor tried to hold the blanket around his body. And the wind eventually gave up. Then the sun started with its rays, very mild, and they increased in strength and as they increased … the traveller felt that the blanket was unnecessary because the blanket is for warmth. And so he decided to relax it, to loosen it, but the rays of the sun became stronger and stronger and eventually he threw it away. So by a gentle method it was possible to get the traveller to discard his blanket. And this is the parable that through peace you will be able to convert, you see, the most determined people, the most committed to the question of violence, and that is the method we should follow.[11]

Love is a powerful response to wrongdoing too

I am grateful for the work of philosopher Myisha Cherry in helping me to recognize that my love is a powerful force for justice, especially when it works hand in hand with my grief and anger at wrongdoing. She has helped me to understand that just as we feel multiple forms of love—passionate love, maternal love, love of sunshine—we feel different kinds of anger—anger at wrongdoing, anger at neglect, anger at injustice. Love that seeks to control or dominate is wrong, just as self-serving anger that is focused on payback, revenge or entitlement is destructive. But that doesn't mean that love or anger per se is bad.

I continue to learn inspiring ways to blend anger at injustice with love and hope to change systems from Cindy

Blackstock. She is a member of the Gitxsan First Nation with thirty-five years of social work experience in child protection and Indigenous children's rights. As with many mentors in my life, we've never met, yet her tenacious acts of courage—she worked with First Nations colleagues on a long, difficult and ultimately successful human rights challenge to Canada's inequitable provision of child and family services—her respect for the wisdom of children, and her honest vulnerability to act even when she feels daunted galvanize me whenever I find myself wobbling in my commitment to a difficult cause. In a brilliant interview with Pam Palmater, Cindy shares her commitment to act from a place of love:

> I live near Ottawa and I see lots of protests at Parliament Hill. People are angry and crying. I don't want to be a part of that. What the kids have really taught me is that the most sustainable movement has to be based on love... When I read letters from children, they will say, *What you are doing is wrong, government. Why are you giving these kids dirty water that makes them sick?* But they always sign their letters *"Love so-and-so"*... We need to see unfairness wherever it exists and we need to stand up for not just First Nations kids, but for all people who are being treated unfairly. That's the type of role modeling we want, to co-create a society where no one is left behind. If you are like me and you don't feel you are smart enough, just remember... we need to show our kids that even we are prepared to address our fears and insecurities because we love them enough to try.[12]

We are naturally oriented to fairness, not inequality

Betrayal is especially painful because it goes against the innate expectation of fairness you have carried within you since you were a baby. According to developmental psychology researchers, within their first year of life, infants expect things to be distributed equally. They also prefer people who behave fairly over those whose actions are unfair.[13]

Humans have a strong sense of equity, and so do a number of other species. Inequality is not inevitable, nor is it the norm in nature, as evidenced by a 2023 analysis of data from sixty-six species of mammals. Indeed, the evolution of fairness has played a big role in the evolution of mammalian species. According to a summary of the study from UCLA:

> Some mammals share food with relatives who are unable to find their own food, or to strengthen social ties. For example, vampire bats share blood meals with relatives who are weak from hunger, while chimpanzees share meat from a successful hunt with the entire group. Some animals, such as elephants and lions, adopt orphaned young and raise them as their own. These practices mitigate the unequal distribution of resources within a hierarchy and promote social relationships based on sharing, not dominance.[14]

Scientists at the German Primate Center use the term "inequality aversion" to describe monkeys' feelings of frustration when they are treated unfairly by a person. In a poignant study of long-tailed macaques (*Macaca fascicularis*), the monkeys were willing to accept an unfair reward from an automatic feeder, but they reacted with disappointment and frustration when they were given an unfair reward by a human experimenter.[15] They rightfully expected the person to behave fairly.

Inspiring studies like these can help to remind you that the kind of extreme inequality that exists in the world today is *not* natural. It is constructed. Which means it can be deconstructed. That is where I situate hope. As Grace Blakeley puts it, "Inequality is sustained by the unfounded belief that those at the top are destined to rule, while those at the bottom are destined to obey."[16] Illuminating those baseless beliefs opens up more possibilities to reject and change them.

Dismantling global systems of injustice

Rummaging through my notebook, I come upon an entry I wrote in June 2024.

> *I'm thinking about what I need to learn at this moment. Right now, someone I love is in an activist encampment on a university campus that has a high chance of being taken apart by the police which means she's at risk of violence and arrest. I'm learning how to sit in love and support of her and everyone she is with and the justice they're calling for, while not knowing what's happening to my wonderful girl. I find this hard.*

The entry refers to one of the more than 170 pro-Palestine movement encampments that were established in the first semester of 2024, mainly in university settings, across thirty-five countries worldwide. The encampments emerged as part of a movement of overwhelmingly peaceful assemblies held in more than sixty countries. "This massive civic mobilisation, which lasted months and is unprecedented in recent history, emerged amidst severe restrictions on the right to freedom of peaceful assembly at a global level," reports the Office of the United Nations High Commissioner for Human Rights (OHCHR).[17]

The protests recognize the social justice interconnections between climate change, human rights, peace and environmental issues. They seek to uplift anti-racist, decolonial and inclusive practices. The chants of "No climate justice without human rights. No peace without justice" are a clarion call for the widespread systemic transformation we need to achieve climate justice.

These actions are part of a long line of meaningful campaigns that stretches from the divestment movement in the 1970s and 1980s (which demanded institutions drop their direct investments in corporations that operated in South Africa, in order to put international pressure on the apartheid government) through the climate justice marches of 2019.[18] They are fueled by a kind of inclusive anger, or "Lordean rage," as Myisha Cherry dubs it, in honor of an essay by Black feminist poet, writer and activist Audre Lorde. This rage aims to dismantle structures of oppression, not by eliminating the other but through changes in beliefs, expectations, policies and behaviors.[19]

I find hope in the millions of people all over the world who gather in solidarity against the financing of human and environmental destruction. "Divestment movements have developed in many contexts, ranging from the genocide in Darfur and Israel's policies vis-à-vis the Palestinians to the tobacco industry," writes Joshua A. Schwartz, an expert in international relations at Carnegie Mellon University. "The most prominent contemporary divestment movement is the campaign to divest from fossil fuel companies.

Currently, over 1,500 entities controlling almost $40 *trillion* in assets—such as the Irish government, New York City, The Rockefeller Foundation, Harvard University, and the World Council of Churches—are part of this movement."[20]

Thanks to the climate strikes led by children and youth, one in every eight people on the planet now lives in a jurisdiction that has declared a climate emergency. This, too, is hopeful, because climate emergency declarations give rise to action plans that call for specific outcomes, such as zero carbon emissions by 2030. These pledges are valuable devices for holding governments, corporations and other entities accountable for action at civic, regional, national and international scales. At the same time, new organizations and initiatives are emerging to monitor pledges for greenwashing. The Global Investigative Journalism Network, for instance, has a reporting guide on "Holding Governments Accountable for Climate Change Pledges" (see resource list on page 273).

The telltale signs of institutional betrayal

The fact that many people are now talking about "climate justice" rather than "climate change" is a testament to the youth-led movement that demanded—and succeeded in convincing the UN and other international institutions—that climate change be recognized as the social justice issue that it is.

But at what cost? What are the "felt experiences" of students engaged in peaceful protest only to have their own universities attempt to delegitimize them? Where is the justice in unleashing police violence upon them when they are peacefully gathered in support of a more just world?

Anger. Fear. Outrage. Shock. Defiance. Disillusionment. When the institutions you depend on cause you harm, the psychological distress and long-term relationship costs are profound. Jennifer Freyd, professor of psychology at the University of Oregon, introduced the term "institutional betrayal" to describe the "wrongdoings perpetrated by an institution upon individuals dependent on that institution, including failure to prevent or respond supportively."[21]

Institutional betrayal harms in painful and destructive ways. Avoidable government failures in managing COVID-19, for example, caused unnecessary deaths, economic ruin, increased inequality, psychological and emotional distress, and a loss of trust in health services that has ongoing implications for public health. The same holds true for the "slow violence" associated with climate change.

Research reveals that perpetrators of wrongdoing display a specific pernicious pattern in response to being held accountable for their behavior, whether it be incidents of partner abuse, sexual violence or covering up toxic waste spills. The pattern is Deny, Attack, and Reverse Victim and Offender (DARVO).[22] First the person or institution denies any wrongdoing (*Not here!*). Next they attack the credibility of the person who has been harmed (*You're making it up*) and gaslight them (*You can't be remembering it right*). Then they frame themselves as the victim in the situation, and the real victim as the "bad one" (*How could you do this to me?*). If you've ever dealt with someone who refuses to accept or make amends for the harm they've done, you'll likely be familiar with this pattern.

Support institutional courage

I find hope in institutions that display courage. Instead of engaging in DARVO tactics, they engage in moral action, value transparency and seek the truth, even when it's difficult, costly or a threat to their reputation. The Ocean Cleanup is a nonprofit organization involved in developing and scaling technologies to rid the world's oceans of plastic. I have been following its work since Boyan Slat, a Dutch inventor, founded it in 2013, when he was eighteen years old. The team now consists of 145 engineers, scientists, researchers, modelers and other professionals. It has been highly successful in its very challenging work. What I find courageous about them is not only the scale of the task they are engaged in but also their ongoing commitment to share both their successes and their failures. They are courageous and transparent about what works and what doesn't.

The OHCHR 2024 report is courageous in denouncing the failure of universities to protect their students' rights:

> At a time when apathy and mistrust increasingly dominates policy-making spaces... the fact that hundreds of thousands of diverse students worldwide are exercising their academic and fundamental freedoms to advocate for collective change, to protect human rights and end atrocity crimes and promote peace, should be encouraged and facilitated, not condemned, silenced, repressed or criminalised.[23]

The Center for Institutional Courage has great resources to help organizations and institutions promote institutional courage and build it into their policies and procedures.[24] The following checklist is inspired by their work.

Check for institutional courage

I use this checklist with students in the classes I teach to assess how courageously the places where they work or study respond when a problem arises.

1. Were efforts made to prevent this issue from occurring in the first place?

2. Is it easy to report when something goes wrong?

3. Could you disclose the issue to a confidential and neutral third party?

4. Did others who knew about the issue believe you and stand in solidarity with you?

5. Did you feel safe and heard?

6. Was the issue covered up?

7. Were procedures or policies changed to address the issue?

8. Did you receive the care you needed to heal from what happened?

9. Did you receive an apology?

10. Are you satisfied that appropriate measures were taken to resolve the issue, ensure your safety and prevent a recurrence?

Welcoming different
and confronting perspectives

For the past few years, my daughter Esmé and I have been in a challenging and meaningful conversation about peace, violence, social transformation and activism. At the same time, I've been engaged in an ongoing dialogue about hope with my friend Mavis Underwood of Tsawout Community, W̱SÁNEĆ Nation. These conversations with women I admire who are respectively in their early twenties and seventies are rich in genuine caring, personal perspectives, acts of courage and scholarly thought. Through their activism and their generosity, I've encountered ideas that were difficult for me to grasp, let alone consider.

It was Mavis who introduced me to Dian Million's concept of "felt theory." Throughout her inspiring scholarly work of the same title, Dian demonstrates the power of embracing emotional knowledge and shows how often this type of knowledge has been silenced by discrediting its worth. Referencing writers like Maria Campbell, Lee Maracle and Ruby Slipperjack, she writes that Indigenous women

> participated in creating new language for communities to address the real multilayered facets of their histories and concerns by insisting on the inclusion of our lived experience, rich with emotional knowledges, of what pain and grief and hope meant or mean now in our pasts and futures... Our felt scholarship continues to be segregated as a "feminine" experience, as polemic, or at worst as not knowledge at all.[25]

It was Esmé who helped me to recognize the need to challenge my naive belief that peace is a universal good.

Through her, I encountered the concept of "weaponizing peace" as described by Yuvraj Joshi, a professor at Brooklyn Law School. "Racial justice opponents regularly wield a purported desire for peace, stability, and harmony as a weapon to hinder movement toward racial equality,"[26] he writes.

Maintaining the status quo is not value neutral. "Stability" in an unequal world is built on the injustice, despair and frustration of marginalized communities and other species, and should be seen for what it is—ongoing systemic violence. "Inequality is a part of the foundations of capitalism," writes Grace Blakeley in a 2024 publication of the London School of Economics. "For a capitalist economy to function, some people must own all the resources required to produce commodities, while others are forced to sell their labour power merely to survive."[27]

The pervasive belief that you or I or anyone else makes their wealth on their own leaves out the larger context. All wealth in a capitalist system involves, at some level, the exploitation of "common resources"—which might be entire ecosystems—and the labor power of human and other-than-human beings.

Yet day-to-day life in popular culture proceeds as if this vastly unfair and destructive system doesn't exist. When we decide not to see what is hidden in plain sight, it creates apathy in the form of an "inability to grasp, fathom, and sympathize with the plight of those who are forced to endure those conditions," writes Jan Slaby, a philosopher at Free University Berlin.[28] Martin Luther King Jr. described this as an "obnoxious negative peace": an illusion of peace that must be disrupted to achieve long-term positive peace.

Those of us who do see the enormity of the nonresponse to something so urgently in need of global-scale action feel additional anxiety and rage about the nonresponse, on top of our reactions to the issue itself.

The more I understand what is happening on the planet from the perspective of inequity and betrayal, the more focused I have become on the necessity of repair. As philosopher Ben Almassi expresses so beautifully, "Environmental injustice and other wrongs do moral damage to our relationships—intercultural, intergenerational, interspecies, otherwise—which calls for processes of moral repair if these ecologically significant relationships are to be made healthier than they have become in the wake of wrongdoing."[29]

Treat climate justice as a reparative quest

For all of these reasons, I am drawn to the concept of a reparative quest. I am thankful to Pumla Gobodo-Madikizela, a clinical psychologist working in the area of transgenerational trauma, forgiveness and transformation, who originally conceived of this concept. She is the founding director of Stellenbosch University's Centre for the Study of the Afterlife of Violence and the Reparative Quest. She developed her idea of a reparative quest in response to apartheid, the institutionalized system of racial segregation that persisted in South Africa from 1948 through the early 1990s.

After the end of apartheid, the Truth and Reconciliation Commission (TRC) invited victims of gross human rights violations to give statements. "One of the most important lessons from the experience of being on the TRC and listening to people's stories is that this is the moment that the real reckoning with the past begins," she says. In an interview hosted by the Templeton Foundation, she describes the experience of witnessing the testimony of former police colonel Eugene de Kock, nicknamed "Prime Evil" for his role in killing and torturing anti-apartheid activists. "The widows of men that he had murdered... reached out with forgiveness to him," she said. "I had no reference points for... forgiving someone who murdered your husband in this violent way."[30]

Pumla was compelled to try to understand how forgiveness could happen in such impossible circumstances. Through her conversations with the women, she discovered their commitment to something much larger than self. There was a genuine desire for the offender to change. "There is a lesson for us to understand about the value of these moments of encounter," she says. "What they can teach us about what's possible, even in the face of so much tragedy. It has led me to think about the process that is unfolding as a process of repair." Forgiveness is about bringing something into the world so something changes in the world.

I find this conception of repair in the midst of continuing struggles incredibly compelling. It speaks to the extraordinary capacity to choose to act on behalf of the greater good. And it recognizes that the work of societal transformation is ongoing and in constant need of repair. Thirty years after apartheid, South Africa continues to grapple with a legacy of racism and segregation.

Reconciliation is "a movement of hope," writes Karine Duhamel, an Anishinaabe historian and a member of Opwaaganasiniing (Red Rock Indian Band) in northwestern Ontario. Karine served as director of research for the National Inquiry Into Missing and Murdered Indigenous Women and Girls in Canada. "This is not to deny that there is an ugly, wicked and horrific past that society, including governments and citizens, need to address," she says. "Colonization lives on in many ways, and the damage can only begin to be undone by thinking forward. In short, acknowledging the past within the framework of hope means we can imagine a way forward; it means, in very simple terms, that the solutions we are seeking become available, once we stop feeling trapped in the guilt of the past and begin imagining a better future for us all."[31]

Truth and reconciliation processes inspire me to imagine how the Earth and the 8.7 million species, including us humans, who live upon it might heal from the ravages of mass-scale habitat destruction, species extinctions and the devastation of climate change. How do you envision yourself within this reparative quest?

The stance *I am on a reparative quest of transformation* is an invitation to co-create that answer. It's an invitation to use hope as a driver of social change in ways that nurture, heal and care for yourself, each other and the greater-than-human world. For me, it involves:

- Speaking the truth of global injustice;

- Unlearning personal and societal narratives that hold me in broken systems;

- Amplifying proven solutions;

- Listening with compassion and empathy to people's climate emotions and experiences;

- Considering forgiveness;

- Centering other species.

Unlearn

Questioning your beliefs and immersing yourself in emerging narratives as a regular habit opens you to look anew at the systems that hold the status quo, and your part (consciously or without intention) in reinforcing them. Being vulnerable is a means of opening yourself to being transformed.

To be honest, I didn't always appreciate this as much as I do now. It was when I sat down to write, and discovered I was wrestling with a book that refused to be written, that I slowly realized that many of my ideas were being actively reformed. It is hard to write an integrated synthesis of ideas when you are seeking whole new ways of understanding. I could tell I was really challenging my underlying narratives, and not just updating facts, because I felt insecure and shy, full of quandaries about which ideas about how the world works to carry forward and which to leave behind.

I wrote these pages in the midst of an ongoing struggle to unlearn what the culture of modernity has taught me *should be*. I am grateful to scholar and activist Vanessa Andreotti for her description of modernity as "a single story of progress, development and civilization that is expiring." In those few words, she elegantly encapsulates both the global dominance of this powerful narrative and the reality of its unraveling.

I am unraveling too. I have understood for many years that rampant consumerism, endless economic growth and other driving forces of modernity are at the heart of the climate crisis. Yet for all that knowing, I keep being shocked each time I recognize a new instance in which I've failed to see how my actions or assumptions feed the status quo. I am aware of this tension playing out in the pages of this book. Undoing ways of thinking, feeling and being that render me complicit in perpetuating modernity's systemic injustices, and showing up differently to myself and others (of all species) is essential, super difficult and humbling.

Through her book *Hospicing Modernity*, Vanessa enables me to see the massive societal upheaval in which we are collectively entangled through the tender, catastrophic lens of death—and birth. "Hospicing is about giving palliative care to a dying modernity while offering prenatal care to systems that could be in gestation," she explains in a YouTube recording of a talk she gave to the Joseph Rowntree Foundation. "It depends on how we hospice the old in order to be able to welcome the new. The wisdom of the new system depends on our ability to be taught by the mistakes of the old."[32]

Death, birth, growth—what is true for global narratives is true for ourselves. I appreciate the way essayist Maria Popova celebrates this feeling: "If we don't continually outgrow ourselves, if we don't wince a little at our former ideas, ideals, and beliefs," she writes, "we ossify and perish."[33]

Personally, I struggle to imagine a world beyond capitalism. It's so deeply embedded in the way the world operates, I'm constantly being reinforced to see it as a given. I am committed to keep checking the expiry date on my thinking about capitalism and to challenge it as a forever norm.

What I'm coming to learn is that although economic systems have been around for more than six thousand years, the capitalist societies we live in are relatively young. Writing in the *Oxford Review of Economic Policy*, economist Jeffrey Sachs of Columbia University says:

Modern capitalism emerged in the early nineteenth century in western Europe and the European offshoots of the Americas and Oceania. Recognizing the unparalleled dynamism of the new socio-economic system, Marx and Engels predicted in 1848 that capitalism would spread to the entire world. By the end of the twentieth century, that prediction was confirmed: capitalism had indeed become global, but only after a tortuous and violent course of institutional change in many parts of the world.[34]

Seeing capitalism as a young, albeit globally dominant system opens me up to seeing it as something that could change. It gives me the freedom to look at alternative systems, such as a degrowth economy movement that is emerging beyond capitalism.

Degrowth is an ideology which rejects the assumption that we can have exponential economic growth on a finite planet. Degrowth recognizes that the climate crisis is being driven overwhelmingly by rich countries using too many resources and too much energy. "Who's driving the ecological crisis?" asks Jason Hickel, a professor at the Institute for Environmental Science and Technology at the Autonomous University of Barcelona in an interview for *Resilience*:

It is overwhelmingly the rich countries of the Global North: the United States, Canada, Europe, Israel, Australia, New Zealand and Japan. These countries are collectively responsible for 92 per cent of excess emissions. They have colonised the atmospheric commons for their own enrichment. Meanwhile the entirety of the Global South—all of Asia, Africa, Latin America—is responsible for only 8 per cent, and that's from just a small number of countries... Degrowth is a call to liberate the South from imperial appropriation and decolonise the atmosphere.[35]

Conceptions of degrowth vary with respect to how politics and economies should be reorganized.[36] Yet there is widespread agreement that to avoid the worst possible climate futures, it means less material production and consumption, particularly in the Global North. At the same time, according to Giorgos Kallis, coauthor of *The Case for Degrowth,* it means an increased role for the commons, collectivism, localism, wealth redistribution, meaningful social participation, and reconnection to ourselves, each other and the greater-than-human world. Degrowth principles underpin the volunteerism and community-based initiatives of the Sarvodaya Shramadana Movement, which seeks a no-poverty, no-affluence society in Sri Lanka. We see evidence of them in Finland's forays into universal basic income programs, in which people receive an unconditional income that is not tied to a means test or a requirement to work.

I am drawn to the notion of *frugal abundance* and the way it invites us to imagine and create societies in which everyone has a good life, consumption is low enough to achieve global ecological and social justice, and the material needs of everyone are satisfied.

Embrace uncertainty and stay in relationship

People have beautiful ways of expressing the discomfort that accompanies big shifts in learning and being. In his letters, the romantic poet John Keats writes about beauty, imagination and the concept of *negative capability*—the willingness to make peace with ambiguity, live with mystery and embrace uncertainty. Harvard psychologist Daniel Gilbert explains it more succinctly:

Human beings are works in progress that mistakenly think they're finished. The person you are right now is as transient, as fleeting and as temporary as all the people you've ever been. The one constant in our life is change.[37]

Thinking in new ways and making meaningful shifts around climate justice and other environmental issues is rubbing off on other areas of my life. For instance, I'm quicker to catch myself when I assume how someone is going to respond, rather than checking with them directly.

I find it difficult to remain in relationship when someone is being untruthful or only looking out for their own interests. Yet the hard work of dismantling unfairness requires me to stay in messy disagreements and power-based conflicts. Knowing that I am on a reparative quest helps me to remain in the dialogue as an act of solidarity.

Not long ago I met Nicole Furlonge. She is a professor at Columbia University and the author of *Race Sounds: The Art of Listening in African American Literature.* She helped me to understand that listening does not require agreement. Freed from my belief that *if I listened, I was somehow condoning something I didn't agree with*, I'm noticing how much more able I am to approach conflict with agility, even creativity; to dampen the polarized positions that escalate divides. When I choose to truly listen, listening gets us a little closer. Instead of running, I choose to honor my emotions *and* stay in, because distancing only serves to perpetuate the broken status quo. It's true what Nicole told me: "Listening helps you to develop conflict intelligence."

Nicole shared this short practice that I find quite helpful to remind myself who I am and who else I am bringing into whatever situation I'm in. It's a lovely one to share with a friend.

Listen to who you are

Working silently for two minutes, write down your answers to these two questions:

Who are you from?

Who do you carry with you?

Then take turns sharing and listening and discussing what emerges.

See your identities as ongoing choices, not forever labels

It is not only broken systems that need transforming; it is the narratives we embody within ourselves that cause us, often unintentionally, to be complicit with or actively reproducing those broken systems.

Being open to being transformed can feel disorienting and uncomfortable. Indeed, losing your confidence, feeling like you honestly don't know what to say or what you think—these feelings and more can happen when you first begin to realize how often what you just automatically say or think is reflecting narratives you didn't even know you had embodied. Some of these narratives have enshrined values and beliefs you may actually abhor.

Each of us has blind spots when it comes to how we see and operate in the world. The ability to spot what serves and what is broken, and to attempt to change these perspectives within yourself, depends on your capacity to recognize your individual circumstances and how your social position and power influence your privileges, experiences and expertise—what has come to be known as your "positionality."[38]

There are lots of resources and workshops to help you explore your positionality. As you do so, try to see the multiple identities that make up *you*—not as static labels, but as something you earn each day through your ongoing actions. I don't believe that I get to claim I love someone, for example, if I am not actively making loving choices in the way I interact with them. In the same way, I can't claim "anti-racist" as an identity. Whether or not I am anti-racist depends on the choices I make (or don't make) each day.

Seeing your identities as ongoing choices, rather than forever labels, enables you to move beyond becoming complacent or disempowered within a particular positionality. Some of the students in the classes I teach at the University of Victoria in British Columbia, for example, introduce themselves through a series of positionality statements, such as "I am a white, cisgender, female settler from the traditional territory of lək̓ʷəŋən-speaking and W̱SÁNEĆ peoples." As Emma Battell Lowman and Adam J. Barker describe, "Active identification as a *Settler Canadian* can signal to others that we are ready and committed to honestly addressing settler colonialism in Canada."[39]

Identifying yourself as a settler is valuable, *and* it is just one step in an ongoing process of personal decolonization and regeneration. Don't stop at the label. What you do each day with your awareness is what's important. "Settler allies are made, not self-proclaimed," writes Tricia McGuire-Adams. "As many Indigenous peoples continue to identify ongoing racism, there is a need for informed, unsettled, anti-racist allies willing to challenge their own complicity to then take action when anti-Indigenous racism occurs. Actions include critical self-reflection, confronting white supremacy and implementing demonstrably anti-racist acts."[40]

Swim in unfamiliar waters

Make sure you are not putting it on others to teach you rather than taking your own responsibility to learn. I find it helpful to practice this by thinking of something I learned from the animal care staff at the Monterey Bay Aquarium. When a fish needs special care, they don't treat the individual, they treat the water in which the fish lives. Immerse yourself in the cultures in which you swim. I intentionally work in diverse collaborations. I subscribe to online newsletters, like Reimagined (formerly Anti-Racism Daily), which invests in BIPOC writers and helps me to develop terminology and strategies to dismantle white supremacy. I follow the blogs of the Indigenous Leadership Initiative, which spotlights developments in Indigenous-led conservation in Canada, and I listen to podcasts like *Crossing the River,* a project of the More Than Human Life (MOTH) project at New York University School of Law that provokes ways of thinking about how to listen to the voices of the more-than-human world.

I often turn to other species to push me beyond my narrow human-centric ways of thinking. I find crab molting, for example, an inspiring metaphor for personal growth. Unlike yours and mine, the skeleton of a crab is on the outside of their body. It takes the form of an exterior shell often consisting of separate plates of a protein called chitin and calcium carbonate. A thin membrane creates "joints" which allow the crab to move easily. Each time the crab outgrows their existing skeleton, they literally extract themself, limb by limb, gill by gill, from the too-small shell. The crab emerges as a delicate, soft-bodied being. In other words, a crab must make themself vulnerable time and time again, in order to grow.

A crab grows a new, soft exoskeleton inside the existing one, and then sheds the outer one. After escaping the too-small shell, the crab must keep themself safe while they await the emergence of a new, better-fitting version. During that time the crab fills their tissues with water, swelling to an even larger size so that the new shell that is hardening will have plenty of room to expand into.

We are always outgrowing and becoming anew. The following is a practice I use to help me embrace this ongoing transformation.

Consider what you have outgrown.
Ask, *What are you growing into?*

1. Think about behaviors or beliefs you've been living within. Are there any that you have truly outgrown? What no longer fits the person you are becoming? (I sometimes experience this as a feeling of getting on my own nerves; like the way I am interacting with the world is pinching or rubbing up against myself.)

2. Plan for how you will keep yourself safe during the vulnerable time of change. Remember, growing new ways of being can be risky and can leave you feeling sensitive and exposed.

3. What new shape is forming around you? What support do you have in place to help you grow into the you who is emerging?

Acknowledge guilt
and welcome hope

Guilt mixed with inadequacy often weighs heavy on me. I imagine the same might be true for you. I think guilt is so present because no matter what the issue, we are constantly trying to decide on the best course of action within global economic and political systems that are inherently unjust and ecologically destructive. The scale of climate justice and other environmental issues can create an oppressive sense of futility. Self-loathing is anxious to creep in, fueled by the certitude that someone smarter or braver or more committed would have already contributed so much more of value to ease these very wrong things.

Sometimes people say, "I don't want to feel guilty anymore. Why should I feel guilty for what our ancestors did in the past?" I understand why it can be hard to embrace guilt as the helpful emotion it is. What I am learning is that for me, guilt goes hand in hand with the profound shifts necessary to name and unlearn my privileges, just as humiliation is part of realizing the depth of my ignorance. I can't engage in transformation in any meaningful or truthful way unless I'm willing to sit with these confronting feelings. When I do so from my stance of hope, I find myself more able to find my agency rather than veering off into disempowering self-loathing.

I think this focus on hope as a means to channel guilt in a generative direction is what Karine Duhamel is advocating, particularly with respect to reconciliation. Resist the unhelpful tendency to get caught up in guilt or let it slide into self-flagellation in which larger issues become all about you and your own feelings. "Guilt rarely moves people forward; in fact, it often traps them in the past, stuck in a feeling that they just can't shake," she writes in an online article published by the Canadian Museum for Human Rights. "Hope that emerges in a proactive way and still understands and knows the past, this kind of hope implies action. It necessarily involves the will to get there and it forces us to consider how we will do so."[41]

❧

Show up authentically

Examining how climate change interacts with power structures to create and reinforce power, privilege, disadvantage and discrimination is essential and an important part of bringing an intersectional lens to the ways we understand social-environmental injustices.[42] Leah Thomas and other climate activists use the term "intersectional environmentalism" to advocate for "an inclusive form of environmentalism that advocates for the protection of people as well as the planet."[43]

Learn to recognize the power imbalances that exist all around you. Reflect on ways you show up in solidarity. Consider personal choices such as working for companies or organizations whose environmental or social justice values align with your own, or if you are in a personal position to do so, "climate quitting" those that don't. Expect yourself to bravely stand up for climate justice, especially when it may be socially awkward or politically contentious to do so.

"Let's not stop at self-reflection," writes Momtaza Mehri, the Young People's Poet Laureate for London and an independent researcher.

> Unlearning personal prejudices should coincide with undoing the structures, logics and economic arrangements ... Being courageous enough to reimagine the world as we know it will only deepen our genuine solidarity with those who are currently struggling to survive it. Instead of timidly admitting to our various privileges, let's ask ourselves what a world where all black life matters everywhere would look like—and accept nothing else.[44]

It's hard to learn what needs unlearning in your life. Give yourself grace as you welcome the uncertainty that accompanies deep learning, and show up as the person you are becoming.

Anchor your activism in your unique interests

When you understand not only how you feel about climate change (for example), but why, from your particular social position, you feel the ways you do, it can open up your awareness of new social and political pathways for action. Oil sands workers, for instance, often feel pulled between their concern about climate change and the need to maintain their jobs in an industry that is a major contributor of greenhouse gas emissions. Recognizing that many oil sands workers feel passionate about climate change inspired the creation of Iron & Earth, an initiative led by fossil fuel workers. Its Net-Zero Pathways program provides training, upskilling and internship opportunities in the renewable energy job sector and connects renewable energy employers with skilled workers.

There are lots of helpful lists to let you know the most impactful things you can do to advance climate justice. I value how Project Drawdown, for example, uses big data analysis to inform their recommendations. Their list has many suggestions, including reducing food waste; eating a plant-based diet; using renewable energy; and getting around greener by biking, walking, using public transit or car shares. I appreciate their proviso that their list focuses on solutions people in high-income countries can directly implement, since it's these people who, while being a minority of the global population, are responsible for at least 90 percent of excess global emissions.[45] Their Drawdown Roadmap focuses on collective scales of change, providing videos and other resources to highlight the most important climate actions governments, businesses, community organizations and other groups can take.

In addition to these valuable general actions, consider how to weave climate actions into your unique passions or daily work. Many people, for instance, are part of a global upsurge of religiously inspired ecological concern.

Engaging from faith

Every major religion on Earth has made climate change action declarations. (The Yale Forum on Religion and Ecology has a comprehensive list.[46]) In 2023, high-level faith leaders from around the world gathered at the Global Faith Leaders Summit, organized by the Muslim Council of Elders, to sign a statement of support for urgent climate action.[47] Pope Francis's encyclical *Laudato sí* explicitly addresses and argues for the connection between the climate crisis and socioeconomic inequality. Indeed, 2025 is a jubilee year (something that happens every twenty-five years) in the Catholic tradition, and the theme for Jubilee 2025 is "Pilgrims of Hope": "a year of hope for a world suffering the impacts of war, the ongoing effects of the COVID-19 pandemic, and the climate crisis."[48]

I am not religious *and* I'm much persuaded by the words of Larry Rasmussen, a renowned Christian environmental ethicist who argues for an alliance of ecology and spirituality across religions in this time of global crisis. In his book *Earth-Honoring Faith* he writes that science is indispensable to understanding what is happening to our changing planet, but we need a chorus of world faiths to counter the consumerism, utilitarianism, alienation and oppression that has pushed us to the brink. "It is foolish not to tap millennia of fluency in the arts of life instruction and renewal, just as it is foolish to overlook the religious loyalties of some ten thousand religions and 85% of the planet's people," he writes.

You are more likely to stay committed to habits that are connected to things you do on an ongoing basis, particularly when an issue you care about takes a turn for the worse, or you feel disempowered by the failure of a government to act, or any other real and discouraging events. Use the following practice to explore how you might integrate your passions with your activism.

Act from your unique gifts

1. **Ask yourself: What am I passionate about? What do I spend a lot of time doing?** (Sports? Art? Knitting? Pinball? Religious practices? Cycling? Being outside? Travel? Work?)

2. **Look online to find people who are combining your passion with climate justice action.** (Interfaith Power and Light, for instance, brings together people of diverse faiths to act on climate change. Queers for Climate Justice creates spaces within LGBTQ+ events for climate justice organizing. The Craftivist Collective uses craft to drive societal change, particularly among folks who don't see themselves as activists. Planet League mobilizes sports fans to take climate action.)

3. **Consider joining one of these groups or draw inspiration from the tangible ways other people are linking their passions or professions to their climate justice engagement.** (I was excited to meet an anesthesiologist who told me he and his colleagues recognized anesthesia gases as greenhouse gases and had successfully banned the use of the field's highest-emission-producing gas.)

4. **Get creative.** (Climate justice demands multispecies justice. I am inspired by the creative way the Sounds Right initiative is enabling nature to generate royalties from its own sounds to support its own conservation.)

WHILE I HAVE BEEN WRITING this book, all kinds of terrible and wonderful things have happened in the world. For all the reasons we've been exploring in these chapters, it's easy to see the bad. The most impactful contribution you can make to the issues you care about is to be truthful about what is happening by seeing the problems *and* the good too, and to spread what works as far as you possibly can.

Track progress on the issues that matter to you

Isn't it wonderful how often the people you love pop into your mind throughout the day? The joy you experience by truly knowing them; by following what is important to them—their worries, their triumphs, their secret wishes. I love the feeling of keeping track of the folks I love, and I love it when they track what matters to me.

I use the same approach to keep myself immersed in issues that matter to me. It's a really great way to remain engaged in the full complexity of an issue rather than getting buffeted about by a headline popping up out of the blue. When the COVID-19 lockdowns saw the return of single-use plastic bags in many places where they had previously been banned or had a charge for use, for example, I was disappointed, but I wasn't too worried that this turnaround would become the norm. I knew, because I had been tracking it for some years, that the movement away from single-use plastic bags is global in scope. Bans, tariffs and incentives to reduce their use have been in place long enough for the results to speak for themselves. Supermarket plastic-bag charges led to a 98 percent drop in usage in England as of 2023.[49] Botswana and China experienced a 50 percent drop. South Africa, Belgium, Hong Kong and Portugal reduced use by 74 to 90 percent.

There is overwhelming support for banning single-use plastics. As of 2024, for example, 85 percent of people polled across thirty-two countries believe that a global plastic-pollution treaty should outlaw single-use plastics.[50] Knowing this enables me to follow the November 2024 gathering of delegates from 170 countries to negotiate a legally binding global treaty to curb plastic pollution without being thrown by the inevitable doomist headlines that reported the outcome of the talks as a failure. The talks are not a failure. They are ongoing. Setting up a *legally binding global treaty* as something that could be done and dusted in a super-round of talks, and then reporting the outcome as a failure when that unachievable goal is not achieved, doesn't represent a failure of the talks. It represents a failure of journalism. It creates disempowerment.

Choosing to be on a reparative quest means intentionally looking for where positive developments are happening and choosing to invest your energy in helping those to grow. To get you started, the practice on the next pages includes a list of some of the trends I am following that are moving in positive directions, and that have a meaningful impact on climate change and associated biodiversity issues. That doesn't mean they don't backslide and it doesn't mean they are "solved." What it means is that they are developing in directions we need more of.

Track positive trends
like your closest friend

1. Choose an issue that you really care about.

2. Put a note in your calendar to check back at least once every two months to see what is emerging to tackle it.

3. Use words like "progress on _____" or "_____ solutions" or "positive steps toward _____" in your searches. (Expect to have to wade through many problem-oriented articles before you find one that is focused on positive progress around the issue.)

4. Make sure to look across different scales for signs of meaningful change. (Many cities and states or provinces, for example, are far more active than national governments when it comes to plastic bans or net-zero goals.)

5. Remember to look for progress across business, governmental, nongovernmental and other sectors.

6. Expand your search to include actions occurring in a diversity of cultures and countries.

7. Layer other filters onto your search for positive progress. (Cities are responsible for around 60 percent of greenhouse gas emissions. On a rapidly urbanizing planet where more than half of all people live in urban areas, cities are key centers for climate justice action. Progress happening at a city level on this issue thus has a cumulative impact.)

8. Notice how you feel as you begin to see the positive progress that is happening around an issue you care about. (When you follow issues in depth and over time, you're less affected by one negative headline or development—you see the bigger picture of what's working and what still needs to be done.)

9. Reflect on what has most inspired you as you've been tracking progress on this issue over time. Commit to helping a favorite proven solution to grow.

Some of the trends that elin follows:

- Nature-based solutions

- Legislation, bans and tariffs on plastic pollution

- Indigenous-led habitat and cultural restoration

- Urban greenspaces

- Recognition of the resilience of other species

- Plant-based eating

- Just energy transition

- Electrification of transportation

- Rights of other species

- Rewilding and wildlife corridors

- Green finance

- Climate litigation

- Ecological restoration

- Food waste

- Thrifting

- Solutions journalism

- Degrowth economy

- Rights of future generations

The right to transit justice

While commuting on a very full passenger ferry along the Bosphorus waterway in Istanbul, Turkey, I was thinking about cities in the context of mobility justice. It felt liberating to move through the most populous city in Europe with people of all different ages via a comprehensive public transport system that bridges two continents—Europe and Asia—and includes buses, metros, trams, ferries and even a funicular (a type of cabled railway).

Access to public transportation is a justice issue. It has a significant impact on how and where you can live, work and study, collective air quality, public health, global greenhouse gas emissions and much more. Cars are inherently inequitable and inefficient, spending on average 96 percent of their time parked, taking up urban space that could serve more beneficial purposes.[51] It's common practice for youth I know in California to take a year off school in order to work to make enough money to buy a car so they can physically get to a job to make enough money to go to college. In other words, they need a job to afford the personal vehicle to get them to further education or a job.

A look through solutions journalism sources at what's happening around mobility justice reveals lots of inspiring examples that are answering the demand for environmentally sustainable, economically productive, safe, accessible and affordable travel. Mobility is such a complex issue, we'd be hard pressed to label any one of these examples as the perfect solution. Yet all of them are literally and metaphorically moving us closer to where we need to be.

One of the things that I find invigorating about tracking issues like an attentive friend is the thrill of finding real-world examples of something generally accepted to be super difficult that is actually being tackled successfully. Public transit works best in high-density areas. How to serve people living in small cities that buses don't serve is a challenge. In France, for instance, 81 percent of bus passengers are on the two hundred biggest routes.

For the past few years, France has been experimenting with an innovative approach that actually pays drivers of zero- or low-emission vehicles to give rides to others traveling in the same direction. Drivers are paid three euros per passenger and the passenger rides for free. Drivers are also given a bonus of one hundred euros after completing ten rides.[52] This short-distance carpooling innovation joins up the transit system outside of the main transit lines. The program is so popular, 94 percent of new carpooling drivers continue to offer the service, changing their travel behavior for the better. This is very good news for greenhouse gas emissions and for reducing traffic congestion.

Charging cars to enter certain parts of a city (commonly known as congestion pricing) is another way cities have successfully lowered emissions, air pollution and traffic congestion. In 2003, London, England, started charging a premium on cars coming into the City. The revenue was funneled back into improved public transit. Twenty years later, the scheme was expanded to cover all London boroughs. Despite London's population increasing by more than a million people between 2016 and 2024, air pollution concentrations have fallen by 65 percent in central London, 53 percent in inner London and 45 percent in outer London.[53] Roadside levels of nitrogen dioxide pollutants fell by 49 percent.

Here is another case of why it is so important to check the expiry date on what you are sure you know. When the

London plan came into practice, electric cars were exempt from the charge. This incentive, along with the growing popularity of electric cars, has resulted in a situation in which electric cars have become so numerous, they are now causing congestion. In response, London has just changed the congestion pricing to include all cars, including electric vehicles.

Innovative programs like these are valuable lighthouse examples because they make it easier for others to adopt a proven practice. As I'm writing, Delhi is making plans to introduce congestion pricing. Stockholm and Singapore have been doing it for years. Dubai is considering it. (When I was writing this book in the summer of 2024, New York City was wavering. By January 2025, as I prepared the manuscript for publication, NYC was implementing congestion pricing.[54]) That's a wise decision. "Anywhere congestion pricing is put in, it works," says Steve Cohen, a professor at Columbia's School of International and Public Affairs, in a 2024 interview in *The Washington Post*.[55]

Electric scooter sharing on a mass scale in Indonesia, Thailand and the Philippines; the global expansion of dedicated bike lanes as cycling grows; autonomous electric passenger ferries in Stockholm: these are some of the many ways cities are making mobility more joined up, accessible and climate friendly. In 2020, the entire country of Luxembourg made all transit, including all buses, trains and trams, free for everyone, citizens and tourists alike. Boasting the highest per capita number of millionaires in the world, it continues to be attached to its luxury-car culture. And yet the free transit system is still achieving important results. In the quest for "a just world on a safe planet," as the Lancet Planetary Health-Earth Commission's 2024 report puts it,

Luxembourg has succeeded in framing free public transit as a fundamental right.

Don't let distrust rob you of agency

As you are reading about these examples, you may find yourself slipping into cynicism. In the current flurry of disinformation, polarization, greenwashing and public shaming, you don't want to be duped or intentionally misled. You may not feel you have adequate, trustworthy information upon which to base a decision. You may worry that if you choose to act on an issue, you might make matters worse. Perhaps you feel uncertain about taking action in a group with others who are not perfect. (I find Bono's perspective helpful. "You don't have to agree with everyone on everything if the one thing you agree with them on is important enough," he says in a BBC Radio 4 interview.[56])

The practice on the following pages is intended to help focus your decision-making when you're trying to decide what specific climate justice or other environmental actions to take.

Equip yourself to act

1. Keep a growing list of organizations you *do* trust.
I have a few organizations and programs—like the Monterey Bay Aquarium's Seafood Watch program, Project Drawdown, *New Scientist*, the Prairie Climate Center, Our World in Data, Yale Center for Climate Change Communication, the David Suzuki Foundation, the Solutions Journalism Network, MOTH and the Stockholm Resilience Centre—that I know I can count on for reliable advice. (I also exercise the "check the expiry date" practice to make sure that the open, transparent, evidence-informed practices of these organizations continue to be in effect.) Generate your own list and keep adding to it. Steady yourself on these familiar stepping-stones to regain your confidence about what action feels right for you to pursue.

2. Tackle decision paralysis by asking yourself these reflective questions.
The following questions are inspired by the work of the Building Movement Project, a nonprofit that supports organizations in advancing progressive change. They will help you to consider whether or how the actions you are contemplating will align with your values and capabilities and the needs of those you are seeking to help.

- If I choose to take this action, will it make a meaningful difference?

- If so, how will I know?

- Do I have the time, money and capacity to respond effectively?

- What other activities might I have to stop doing in order to commit to this action?

- Who needs this help? Have I asked them what actions would be most useful and welcome?

- Is this work already being done by others? If so, how might I collaborate with them?

- How can I make sure I'm contributing real value and not just duplicating or diminishing other initiatives already in place?

- What conflicts and risks might I encounter?

- How can I minimize them?

- How will I navigate any ethical dilemmas that may arise?

- Can I be kind to myself when I make mistakes?

- Am I prepared to adapt my actions to the situation as it evolves?

- Do I have a community of support to see me through the difficulties I may encounter?

- How will I explain what I am choosing to do to the special people in my life?

- How will I know when it is time for me to end my involvement?

- How will I share what I am learning throughout this experience?

Schedule time to reflect

A few years ago, I was lucky enough to spend time in Finland, where I was introduced to a meeting practice that I have used ever since. At some point in a meeting, someone in the group would suggest "reflection time." The dialogue would stop for half an hour while each person stepped outside for a walk or sat quietly with a coffee, reflecting on ideas that had emerged so far or the best way forward.

I am an extrovert who thrives in collaborative conversations. Yet I was surprised by how helpful it was for me to stop and reflect. I was able to see good ideas that had emerged earlier in the discussion that had been missed, for example, and to check in with not only what I was thinking, but also what I was feeling. Unlike a chatty coffee break, reflection time invites you to focus on the meeting without distraction. What I discovered was how often the post-reflection portion of the meeting benefited by insights that emerged during the personal reflection time. Taking time to reflect individually improved the productivity of our time together. So whether you are planning to act as a group or pondering on your own, give yourself pause to reflect.

The choice to forgive

For me, forgiveness is one of the most difficult and transformative things imaginable. I think that is why I have such awe for and commitment to trying to learn from truth and reconciliation processes. I do not know if the people who find the courage to speak about the harm they have endured have forgiven the perpetrators—whether the perpetrator is an individual or an entire system. The choice to forgive is entirely personal. And yet, its implications are global in their potential impact. Forgiveness is about shattering existing worlds in which these atrocities could exist.

Climate injustice should never have happened, nor should it continue to happen. Choosing to forgive a terrible thing that is still happening is not about condoning or overlooking or minimizing. Forgiveness is about acknowledging the truth of what has happened and is happening, and accepting the reality of the harm that can never be undone. It is about choosing to go forward in the full presence of the truth and the hurt. I am always trying to learn how to do that. I seek to one day have the grace that I see in others when they choose to extend this extraordinary gift of hope.

In my own experience, the most important part of healing a conflict is agreeing on the truth. When the truth is in dispute, meaningful progress is thwarted and suffering is prolonged. Yet if those in conflict can come to a common understanding about what happened, even the most heinous situations can sometimes be reconciled.

Essayist Maria Popova writes lyrically about how forgiveness can deepen and transform our relationships:

> The richest relationships are lifeboats, but they are also submarines that descend to the darkest and most disquieting places, to the unfathomed trenches of the soul where our deepest shames and foibles and vulnerabilities live, where we are less than we would like to be. Forgiveness is the alchemy by which the shame transforms into the honor and privilege of being invited into another's darkness and having them witness your own with the undimmed light of love, of sympathy, of nonjudgmental understanding. Forgiveness is the engine

of buoyancy that keeps the submarine rising again and again toward the light, so that it may become a lifeboat once more.[57]

Dr. Pumla Gobodo-Madikizela inspires me to imagine changes of heart that I thought were impossible. She has witnessed people who have committed terrible atrocities, and the people they have hurt, turning toward each other instead of turning away. Sometimes, forgiveness is a posture we endeavor to adopt, rather than a task we can complete.[58]

Forgiveness is about totally giving up on a different past. I find this a truly powerful idea. As long as I am unwilling to forgive, I am trapped in an impossible search for the thing that happened *not* to have happened. "'No future without forgiveness' is not just a glib, smart slogan," said Desmond Tutu. "It happens to be the truth, whether we are thinking of an intimate relationship between two persons or several within a community or between ethnic groups and nations."[59]

Consider forgiveness

1. Forgiveness is a decision to move forward.

2. You only need to forgive when something bad has happened. You will always, therefore, be forgiving in the midst of something hurtful or unjust.

3. Remember that forgiveness is not the same thing as condoning. You can choose to forgive something that is categorically wrong. Forgiveness does not make what happened any less wrong.

4. Approach forgiveness as an ongoing choice. You can choose to forgive something today, and keep it as an open, ongoing decision. Forgiveness is not a forever decision. It is an open decision you can make and unmake and make again.

5. The choice to forgive is yours to make. Someone else may ask you to forgive, or expect you to forgive, but whether you decide to do that is entirely up to you.

6. Your decision to forgive must come from within yourself. It is not contingent on someone else's behavior. It may be easier to forgive when someone else has apologized or changed, and you may not be open to forgiving until you see that change has happened. Those may be conditions that you must have. Ultimately, though, whether or not *you* forgive is dependent on what you decide within yourself, regardless of what someone else does.

CONTINUED ▶

7. Look for inspiring examples of forgiveness to motivate you. There are lots of variations on forgiveness. Find whichever one feels right for you.

8. Remember, you were born knowing how to forgive. Whatever you practice, you become better at doing.

9. People have forgiven unspeakably terrible things. They chose to do this so they can keep on living—as a strategy for the self to survive, and for a better world to unfold.

I ENCOURAGE YOU to look deeply at the concepts of truth, reconciliation, justice and forgiveness. We are on a reparative quest of transformation because climate change, biodiversity loss, species extinctions, rights abuses of humans and other species and so many other urgent, global issues are perpetrated by unjust systems that operate at the scale of society and are experienced as suffering by individuals— people and other species.

"Forgiving and being reconciled are not about pretending that things are other than they are," wrote Desmond Tutu. "It is not patting one another on the back and turning a blind eye to the wrong. True reconciliation exposes the awfulness, the abuse, the pain, the degradation, the truth. It could even sometimes make things worse. It is a risky undertaking but in the end it is worthwhile, because in the end dealing with the real situation helps to bring real healing. Spurious reconciliation can bring only spurious healing."[60]

The main challenge and significance of forgiveness is that it asks each of us how we can acknowledge the seriousness of the wrongs committed, and at the same time enable the possibility of a new beginning and a joint responsibility for our shared world. That is why I want to end this section with an invitation to forgive. Forgiveness is the ultimate expression of hope. It is a choice that we can decide to make—and continue to make—as a way of moving forward, as a way of healing, as a way of regaining our agency for climate justice: as a commitment to act on behalf of this remarkable planet and each other.

I Am Nature

I CAN'T QUITE REMEMBER when I first pulled a camping mat and sleeping bag under my covered back deck and hunkered down for the night. That makeshift arrangement has evolved into a full-size bed, down duvets, and a secure sleeping space. But I do know why I sleep outside. I do it because I am nature: I am less than whole when separated from the wind, the rain, the scent of earth, the rising crescendo of geese calling as they migrate through the night. I sleep outside because I am less alive within the sterile confines of monochromatic walls and temperature-controlled interiors, no matter how much I admire and appreciate their aesthetic design and beauty. I sleep outside because I cannot stand to miss even a single morning of spring birdsong, the chill of winter on my nose, or the moment when dusk turns to night. I sleep outside because I like the shock of leaving my cozy bed to hurry to the bathroom across a frozen, moonlit night, and the joy of returning to the warmth of my outdoor nest.

Experiencing the dry crack of subzero temperatures, the deluge of endless rain, or relentless blistering heat fuels my awe for other species. I watch a hummingbird rise high above my snow-drenched backyard, moving into a day with no provisions, no backup, utterly dependent on its relationships with other species and our collective agency. "Birds have no pockets," I tell myself. I have turned this wondrous idea over and over again in my mind for many years, like a well-worn river stone, each time encountering its significance anew. That bird's life, and mine, are not isolated entities. We are both only breathing because of ocean plants creating the oxygen in the air and the fish that disperse the seeds of rainforest plants when the Amazon River floods. I would go hungry were it not for the winds that blow the pollen grains of wheat, oats, rice and so many other plants that I eat. The hummingbird's body undergoes a semi-hibernation every night. Its heart rate and temperature may drop dramatically, and still it loses about a tenth of its weight before the morning. This bird would go hungry were it not for the insects that also feast on those same crops.

When I close my eyes and snuggle beneath the covers, I like to imagine the gentle breathing of grizzly bears asleep in their mountain dens, and babies in Nordic countries napping as snowflakes blanket their prams. Like those babies, I am privileged to sleep in a safe place. I have the resources to make conscious, responsive choices about where to position my bed, what I wear, and how many blankets I need. The same is true for Finnish mothers who demonstrate a fluency for reading the northern winter environment and creating calm, peaceful sleep experiences that honors their children as part of nature.[1]

The practice of outdoor sleeping was actively promoted in Finland as an intervention in response to high infant mortality rates in the 1920s. The challenges and advantages became so self-evident that outdoor sleeping for children is now an embedded cultural practice, a norm in Finnish culture.

In recent years, an avalanche of published studies have revealed the physical, mental, emotional and social benefits of time spent outside. The delights reveal themselves.

Sleeping outside is restorative. It promotes relaxation and a contagious sense of well-being. We become more attentive to the needs of each other as we grow more aware of subtle changes in the weather and the seasons. How can I convey the luxurious sensation of air so thick with lavender, I can taste it on my breath, coupled with the heady realization that an individual bumblebee comes to tend those fragrant blossoms each morning, making micro-adjustments to the time it arrives exactly in sync with shifts in the timing of sunrise as spring gives way to summer? "Look," says my daughter, pausing to sniff a clump of dahlias on her way to join me for a movie as the sun sets. We watch as individual bees fly into their own individual flowers, bedding down for the night as the petals close tightly around them.

True, I could also tell tales about rats and blaring sirens and all-night parties, but those stories would be collaged with my growing appreciation of the social and emotional capacities of rats, and the unexpected beauty of listening to my neighbor, a professional musician, practicing her trumpet in the early hours before dawn, unaware that I was outside too, drifting in and out of sleep to Mahler's "Blumine" movement.

I used to be shy about telling people that I sleep outside. But not anymore. As wild swimming, forest bathing, urban birding, outdoor kindergartens and other nature-based movements rapidly gain popularity, so too does momentum for conserving and restoring biodiversity. In the midst of converging crises of biodiversity loss, climate change, pandemics and so many other urgent issues, the key role of nature-based solutions is increasingly being recognized and mainstreamed.

Lately, I've taken to asking friends and colleagues to sleep outside with me on their balconies or porches when I visit them in big cities. It's a quiet gesture; a sleepy form of activism; a way of connecting to the planetary community. I invite them to sleep outside because where we lay our heads is where we belong.

I AM CRAZY IN LOVE with this world of 8.7 million other species. Ever since I was very young, I have been aware that I am not a *me*, but rather, a *we*. I felt then what I know now: My life—and yours—is completely dependent on other species. Countless organisms, big and small, grant us the lives we cherish each day. They digest our food, protect us from infection, influence our moods and so much more. Without the constant generosity of other species, we would have no food to eat, no air to breathe, no fresh water to drink. We are made from the breath of volcanoes.

I find it wondrous to know that we are continuously being remade by the ecologies in which we live, just like this fox so beautifully described by Ramsey Affifi, a former student of mine, now a philosopher at the University of Edinburgh:

> A fox's habits participate in the watershed she inhabits, while her cells participate in the ecology of her organ tissues. Moreover, the ecology of that watershed interacts with the ecology of its adjacent river system, as well as that of the sea it feeds.[2]

These intricate connections fill me with gratitude and awe. I think the fact that so many of us experience existential crises in response to climate change is a testament to love. We are bound within the genius of nature and distraught by the dire threat to our collective existence.

"Can you imagine a world where nature is understood as full of relatives not resources, where inalienable rights are balanced with inalienable responsibilities and where wealth itself is measured not by resource ownership and control, but by the number of good relationships we maintain in the complex and diverse life-systems of this blue green planet?" asks Daniel R. Wildcat, a Yuchi member of the Muscogee Nation of Oklahoma, and professor at the Haskell Indian Nations University in Lawrence, Kansas.[3]

"Yes!" I want to shout out in answer. That is exactly how I experience life on this wondrous planet.

Deborah McGregor, Canada Research Chair in Indigenous Environmental Justice, describes the mutual respect, responsibility and obligations toward nature within Anishinaabek natural law. "Concepts of love, kindness and generosity are not naive ideals in Anishinaabek society," she writes. "They are principles that have enabled us to thrive for millennia, and may in fact prove to be of the utmost relevance in our quest for sustainability."[4]

We are all individuals with unique personalities

When I was a kid lying in the ravine behind our house gazing skyward at the great vees of birds flying south, I didn't see geese as individual animals. They were subsumed into the greater spectacle. We do not see the piccolo player in the parade; we see the marching band. But Canada geese *are* individuals. They mate for life and remain in pairs together throughout the year. The flocks soaring overhead are often parents, their children and extended family relations. Some geese don't migrate every year. Some don't migrate at all. The Canada goose who poops on the lawn may well poop on the same lawn year in, year out.

The crows calling from the overhead wire aren't just any old crows. There's a good chance they're members of a close-knit family group, living communally across five generations. Yearlings and two-year-olds assist their parents, helping to build the nest, clean it and feed their mother while she's sitting on eggs.

We don't usually think of animals in such specific ways. You probably know the names of favorite dogs in your neighborhood, but how many of us recognize the individual gray squirrels that live on those same streets? Yet those squirrels are committed residents. They learn to cross roads more safely by watching you and your neighbors. They "plant" acorns that grow into the oaks that line our streets. They nest with their grandmothers and aunts and children, and when they retire from a day spent foraging for food, they greet each other with a nuzzle of cheek and lip glands that resembles a kiss.[5]

The same individuality exists in plants. I thrill to the growing openness of academics to proclaim that no two tomato, cactus or corn plants are alike. "Individual [sage-brush] plants respond differently to alarm calls, just as individual animals do," explains Rick Karban, professor of entomology at University of California, Davis.[6]

Plants have agency. They are aware of their surroundings, move their organs with a purpose and pursue goals that guide their future activities. Yet the models we use to make predictions about climate change do not take into account the fact that plants are intelligent organisms that actively shape and manipulate their ecological niches. It's time for climate change models to reflect the dynamic capacities of the plants that shape this remarkable planet. As evolutionary scientist František Baluška and botanist Stefano Mancuso write: "Considering plants as active and intelligent agents has therefore profound consequences not just for future climate scenarios but also for understanding mankind's role and position within the Earth's biosphere."[7]

Life is interconnected
in ways that defy imagination

I am in awe of this glorious, resilient entanglement of life. It is my deepest source of hope. No matter how long I live, or how much I learn, I will never get over my reverence for the seemingly infinite complexities that bind us all into this wondrous whole.

How, for example, do tiny golden-winged warblers, each weighing less than an average pair of earrings, migrate 1,500 miles (2,500 km)? And, more wondrous still, how do individual warblers detect tornadoes more than 550 miles (900 km) away—choosing to take evasive action days in advance, while the weather conditions in the places where they are show no apparent changes in temperature, wind speed or atmospheric pressure?

And who would have expected that a nonstop flight *over* the Atlantic Ocean to Mexico would be the fall migration route of choice for many songbird species? "It's a tremendously risky strategy," says researcher Frank La Sorte in an interview for the Cornell Lab of Ornithology's *Living Bird* magazine. "For populations of these [songbird] species to persist there has to be a benefit to transoceanic migration. Some would argue that flying over the ocean is actually safer—no predators, no pathogens, no buildings. You just can't land."[8]

Many do sadly perish. And here again, an unimaginable wonderment: Their tiny bodies turn out to be an important source of protein for... baby tiger sharks! The songbirds' fall migration coincides with a peak in the population of young tiger sharks in the north-central Gulf of Mexico. Marcus Drymon, a fisheries ecologist at Mississippi State University who made this remarkable discovery, thinks that mother tiger sharks have figured out that this is a good place and time to give birth. It lets their young feast on a windfall of nutrients that fall from the sky, before they have fully honed their hunting skills.[9]

WHAT FILLS YOU WITH AWE? What breathless moments of staggering wonderment do you carry safely in your heart? If you were standing here beside me, I would show you the hummingbird nest that blew down in a storm last summer. I would show you the tiny stitches, actual threads of spiderwebs, that the hummingbird has knotted together to form the soft lichen and cat fur (tufts from my neighbor cat, Huxley) into a tiny cup-shaped nest. A hummingbird nest is a piece of fairytale magic.

Allowing yourself to welcome awe is a crucial step in choosing to be hopeful. It is an act of defiance against the destruction and ugliness of an excavator in an urban construction zone ripping apart the very rocks from which the ground is made. Choosing the stance *I am nature* is a decision to claim that which is true. You are part of a thrilling and infinitely complex system of life. When you recognize that you are nature, you realize the resilience that courses through your veins and pulses through the world in which you wander. No matter what happens, life wants to live. The ecosystems of which you are a part are the most exquisite examples of resilience you could ever wish to encounter.

I am telling you this because the narratives of doom that surround you are very powerful. And, in your decision to be hopeful, you are tasking yourself to challenge them, which is very difficult to do. To make things more demanding still, you are doing this difficult thing at the same time as you are trying to be vigilant against colonial mindsets that continue to oppress to this day. As Hadeel Assali, a postdoctoral scholar at Columbia University's Center for Science and Society, says, "Instead of treating the Earth like a precious entity that gives us life, Western colonial legacies operate within a paradigm that assumes they can extract its natural resources as much as they want, and the Earth will regenerate itself."[10]

As you read through this section of the book, I am encouraging you to make brave choices. I am asking you to open your heart to the incredible resilience that exists on this remarkable planet. I am encouraging you to embrace that resilience because that is the way to reject doomist notions that it is already too late to solve the climate justice crisis. The world, and the species that compose it, are so incredibly resilient that to buy into diminishing narratives about species being helpless victims is a travesty that you don't want to be a part of.

At the same time, I know you also don't want to feed the colonial rhetoric that destroying habitats is fine because the Earth will heal itself. "Colonialism was motivated by the promise of plundering the environment and subjugating populations," explains Anuradha Varanasi, writing for the Columbia Climate School. "And the pervasive and persistent institutions of colonialism make it far more challenging to address the climate crisis and implement solutions, especially in a just and equitable way."[11]

Honor the Earth by celebrating its gifts of resilience rather than giving in to the voice in your head that has been taught to downplay that resilience for fear that you will be letting those in power off the hook. Know that you will hold power accountable, and that you can do that while still celebrating and throwing your energy in support of the capacity of ecosystems to heal.

One of the first ways you can practice doing this is by choosing to unlearn colonialist notions of how nature works.

We are in a thrilling renaissance of unlearning about greater-than-human life

"For two centuries, European colonists tore across the world, viewing nature and land as something inert to be conquered and consumed without limits," explains acclaimed Indian writer Amitav Ghosh.[12] This "settler colonial worldview" laid the foundation for the climate crisis, etching a view of nature as nothing more than an object to be used.

It is thrilling to watch this wantonly destructive narrative being actively overthrown in regions such as the Great Bear Rainforest—the largest intact coastal temperate rainforest on Earth—which stretches from northern Vancouver Island to the Alaskan border. Twenty-six First Nations whose ancestors have been living in and maintaining these lands for millennia assert their Indigenous rights over their territories. A new generation of community leaders are over-seeing a wide variety of stewardship programs that protect land and sea, and support youth to reconnect with the language, stories and practices of their culture. As Merrell-Ann Phare, executive director of the Centre for Indigenous Environmental Resources, explains, environmental benefits aren't the explicit goal of these programs: "They are framed to meet other important community needs such as youth empowerment, supporting aboriginal women, etc. Environmental conservation is a means to meeting these goals."[13]

We are in the midst of a rebirth in how we understand plants and other species within the greater-than-human world. It is a refusal to continue to be bound by inaccurate narratives that shoehorned life on earth into colonial and patriarchal norms. It is leading us to reconsider other species—not as helpless victims, but as individuals and populations that are inherently resilient and creative in dealing with changes. The more we understand and value the agency of other species, the more we can support much-needed ecosystem recoveries.

Diversifying beyond the "fathers" of science

At many points in history, the institution of Science, regardless of where in the world it was practiced, was the exclusive domain of white men. "Since its birth around the same time as Europeans began conquering other parts of the world, modern Western science was inextricably entangled with colonialism, especially British imperialism," writes Rohan Deb Roy, professor of South Asian history at the University of Reading. "The legacy of that colonialism still pervades science today."[14]

I find it hopeful, therefore, to see what is happening as the category of *who does science* is increasingly diversifying to include people from different gender identities, cultures, religions, ages, sexual orientations, abilities, histories and more. As this unfolds, the questions scientists ask are becoming more diverse. Science, like everything else, is better when it is inclusive.

For example, Carolus Linneaus, the man widely referred to as the "father of taxonomy," came up with a classification system in which each species is identified by a generic name (genus) and a specific name (species). It's a system still used today. But it's becoming clear that how Linnaeus classified plants was heavily influenced by the sex and gender norms of eighteenth-century Europe. Linnaeus imagined that plants have vaginas and penises, and referred to plant reproductive organs as "husbands" and "wives." Plants without obvious sex organs, such as ferns, fungi and algae, were classified as "cryptogamic"—"plants that marry secretly."

Linnaeus foisted his worldview upon other species, and it turns out to have been a very poor fit. Ninety-four percent of flowering plants, according to the Royal Botanic Garden Sydney, are "nonbinary" and most have bisexual flowers, meaning they have both fertile male and female organs in the same flower.[15] Freeing plants from inappropriate and inaccurate categories imposed by an eighteenth-century narrative opens up our capacities to recognize and support their astonishing diversity.

Let these examples remind you that sex, gender, race, sexuality and other categories are human constructs. This matters not only for how we perceive other species but also because scientific "facts" about the natural world are frequently used to reinforce normative views of how we ought to behave as people. Recognizing that ideas in science have changed and will change again over time brings a much-needed openness to our current societal discussions on gender, sexuality and more.

Examples of layering human constructs onto other species abound. There's a good chance you learned that male songbirds sing in the spring to announce their breeding territories and attract females. The males sing to call the females. The belief in birdsong as an almost exclusively male trait extends far back in history, and is fundamental to the formation of Darwin's theory of how evolution works. The role of females, as he wrote in *The Origin of Species*, was to listen: "Female birds, by selecting, during thousands of generations, the most melodious or beautiful males, according to their standard of beauty, might produce a marked effect."[16]

Incidents of female birds singing have been regularly dismissed as atypical, rare or the result of hormones within particular individuals having gone astray. Yet a massive study across 1,141 songbird species reveals quite the opposite.[17] In two-thirds of these species, female birds also sing. And reconstructive studies of bird ancestry reveal that female birds have been singing since the very first birds existed.

I find this such an astonishing example of the power of narratives to override what we hear, see and think. Think of all the female birds you've watched sitting on a feeder or perched on a branch, singing to the world, in voices you could not hear because you'd been taught that such a thing simply doesn't occur.

These startling revelations came about thanks to the field of animal behavior becoming more diverse. Sixty-eight percent of research studies on female birdsong published in the past twenty years were led by women researchers.[18] Many scientists are now correcting biases and misguided assumptions, recognizing that female birds have been undercounted and overlooked, which undermines not only bird conservation efforts but also fundamental understandings of ecology and evolution.[19]

Meanwhile, the American Ornithological Society is abolishing colonialist bird names in favor of more bird-centric monikers.[20] Many birds all over the world were renamed in the height of colonialism, usually for the white men who claimed to have "discovered" them. In 2024, work began to change names referencing people and to replace them with ones that better describe qualities of the species, such as their color, behaviors, nesting location or the calls that they make. The practice of uprooting narratives and being open to shifting them to right past wrongs is profoundly hopeful.

Treat other animals like they are someone worth knowing

It's pretty amazing what happens when we commit to knowing somebody, whatever species they might be. It warms my heart to know that college students given the opportunity to train chickens as part of a research study came away with a far deeper appreciation of the magnificence of chickens.[21] Before the study began, the majority of students perceived chickens to be slow learners with only basic emotions like fear, hunger and pain. But after training their chickens, they saw chickens as impressive animals who express a breadth of emotions, including boredom, frustration and happiness.

A bevy of animal behavior research further reveals how complex and caring chickens really are. Chickens have more than twenty-four distinct vocalizations, and they communicate and behave in creative ways. They have a sense of numbers and perceive time intervals. They demonstrate self-control and self-awareness. They reason and make logical inferences. They are intelligent and capable of empathy and have distinct personalities. They are cognitively, emotionally and behaviorally complex individuals.[22]

Birds are known for their remarkable attachment to one another. As far back as the turn of the twentieth century, scientists studying bird behavior likened the urge for birds to be together to "a hunger" or "a craving" or "a social pain" that could only be relieved by reunion with a flock.[23] In fact, the social and vocal interactions within a flock trigger the release of opioids within birds' bodies that reinforce togetherness. What must it be like, therefore, to *be* a flock and then suddenly lose the other or others that constitute who *we* are?

After the loss of a flock mate, individual birds increase their social relationships with those who remain, according to scientists at the University of Oxford. They develop tighter bonds within their social networks. Their behavior parallels how some of us, as humans, react to the death of a friend. A study of people's behavior on Facebook reveals that when a mutual friend dies, friends of the deceased become closer and increase their interactions with one another.[24]

I remember hearing an interview on CBC Radio about soothing the isolation of single pet parrots by teaching them to make video calls to other parrots. Though Rébecca Kleinberger of Northeastern University, the coauthor of the study, was quick to say this should never be seen as a substitute for a living, breathing companion, parrots taught to contact their feathered peers with a video call app did indeed show signs of feeling less lonely.[25]

Evidence of "sentience," as scientists call it, underscores what our hearts already know: Nonhuman animals think and feel. It's not only social mammals like gorillas and humpback whales that are known to consciously experience joy, desire, pleasure, fear, pain and other emotions; it's birds and fish and invertebrates such as octopuses too.[26] In 2022, a sweeping review of current research by the Nuffield Council of Bioethics concluded: "There is a growing consensus amongst the scientific research community... that animals possess a wide range of abilities for complex thinking and social behaviours, and experience feelings which matter to them."[27]

Respect the thoughts and feelings of other species

For Peter Godfrey-Smith, an expert on the philosophy of biology and the philosophy of mind at the University of Sydney, these complex behaviors can only be interpreted as indicators of consciousness.[28] Very different brains or nervous systems, it turns out, do the same kind of work that a cerebral cortex does in human consciousness.

This revelation has implications for the treatment of animals, which, in turn, is driving new policies and laws. Legislation in the European Union, New Zealand, the United Kingdom and parts of Australia now recognizes animal sentience. Other places in the world are following suit. Octopuses, lobsters and crabs will receive greater welfare protection in U.K. law following a report from the London

School of Economics based on reviewing more than three hundred scientific studies conveying the animals' capacity to feel pain and distress.[29]

In 2024, prominent scientists signed the New York Declaration on Animal Consciousness. The declaration is supported by a wonderful range of discoveries. Cleaner wrasse can recognize themselves.[30] Young wolverines cavort down snowy mountains on their backs and stomachs in joyful exuberance.[31] Zebra finches dream of singing.[32]

I'm particularly entranced by the capabilities of invertebrates— the majority of animals on Earth, who don't have backbones and are most often overlooked: Cuttlefish remember details of past events.[33] Bees play.[34] Crayfish experience anxiety that can be altered by anti-anxiety drugs.[35] Tiny fruit flies have distinct sleep patterns.[36] My heart thrills to witness the rising appreciation for these spineless wonders. *The Guardian* newspaper, for example, now runs a popular "Invertebrate of the Year" competition that receives global submissions celebrating everybody from myxozoans (animals one-fifth of the width of a human hair) to a half-ton colossal squid.

Building a richer understanding of sentience and consciousness matters because knowing how somebody thinks and feels can change how we care for them. Research published in 2023 on the health of horses, donkeys and mules working in Egypt, Mexico, Pakistan, Senegal, Spain and Portugal, for instance, found that "animals whose owners believed they felt emotions or who had an emotional bond with them, were in significantly better health . . . than those whose owners did not, or who focused on how profitable or useful they were."[37]

Not something. Someone.

That's what I find so inspiring about the Someone Project.[38]
This series of academically informed white papers openly
celebrates farmed animals as individuals, and the individuality they see in one another. "Identity, emotion and
attention are written all over our faces," writes Lori Marino,
a neuroscientist and expert in animal behavior and intelligence, formerly on the faculty of Emory University. The
same is true for many other species. "Sheep are face experts,"
she explains. They are capable of remembering the faces of
fifty different sheep for more than two years.[39] Their ability
to spot familiar faces is on par with yours and mine. Cows
are good at it too. Chickens recognize members of their
social groups and are keenly aware of social hierarchy. Pigs
are skillful at reading the emotions in *our* body postures.
Like me, they prefer to interact when someone is looking at
them, rather than when that person is turned away. Farmed
animals are not something. They are someone.

Beware the meat paradox

Ducks, chickens, pigs, cows, goats, donkeys and other
farmed animals are unique and distinct individuals, yet
openly acknowledging their inner lives has long been for-
bidden territory. We resist the evidence of their awareness,
feelings and concern about their own quality of life because
it threatens our capacity to see them as commodities, as
food. So we have been taught to objectify them. "The discourse of animal production," say linguists writing in the
journal *Poultry Science*, "consistently represent[s] animals as
objects, machines, and resources."[40] Those who enjoy eating

meat but care about animals frequently dissociate "meat" from its animal origins in ways psychologists have labeled the "meat paradox."[41] Most navigate the uncomfortable feeling of dissonance in this love/eat relationship through a variety of cognitive tricks, often without thinking. Once we categorize an animal as "food," it changes the way we consider them. Researchers reveal that people attribute less intelligence, emotion and personality to animals that are eaten. "We don't call the meat the actual name of the animal. We call it pork and beef and bacon," says Hank Rothgerber, a psychologist who studies how motivated reasoning influences our food choices, in a *Scientific American* article.[42]

Industrial animal agriculture involves raising animals in an intensive way at large scales for the lowest possible cost. Almost all ducks, chickens, turkeys, pigs and cows are raised in these conditions. The scale of this issue is so great, it is difficult to grasp. "It's estimated that three-quarters—74%—of land livestock are factory-farmed. That means that at *any given time*, around 23 billion animals are on these farms."[43] (This statistic comes from Our World in Data, a wonderful organization that collates big data to tackle the world's biggest problems by equipping us with evidence that helps us to understand how and why the world is changing.)

If you could look over my shoulder at my laptop screen right now, you'd find it open to a dozen reports published in the past six months by scientists, food producers, animal rights groups and international agencies involved in animal welfare and animal rights. Reading through this information is painful and discouraging. "Virtually all animal lawyers agree that current animal protection law is painfully inadequate and in need of substantive reform," writes Saskia Stucki, a researcher at the Max Planck Institute for

Comparative Public Law and International Law in Germany.[44] According to Helen Cowie, at the University of York, "Cruelty and welfare issues are still as much of a concern today as they were 200 years ago."[45]

FOR ME, ALL OF THIS is almost too terrible and painful to allow myself to even think about. I imagine you know that feeling. Because you care about other species and you care about injustice, you may have taught yourself to respond to such horrors by doing everything you can *not* to feel. You might believe that you are most powerful in confronting the wrongness of this massive issue by silencing your emotions.

Challenge that urge. Gently remind yourself of the wisdom and power of your emotions. Trust that you are more able to engage with this crucially important and painful issue when you allow yourself to feel all of your emotions and then choose to approach this issue from an empowered, hopeful stance.

When I think of a caged goose, I feel heartbroken and furious that an animal born to fly is imprisoned on the ground. Geese are extraordinary flyers. Some, like barheaded geese, accomplish seemingly impossible feats as part of their ordinary lives. During their migration, they actively fly *over* the Himalayas, the world's tallest mountains, at almost 24,000 feet (8,000 meters), in below-freezing temperatures while breathing air with just *one-third* the oxygen available at sea level. As if that weren't jaw-dropping enough, "crossing the Himalayas from India onto the Tibetan plateau also requires bar-headed geese to ascend for many hours, sustaining the longest sustained rates of climbing flight recorded to date, and possibly even into headwinds," according to biologist Graham Scott and his colleagues.[46]

Bar-headed geese also *choose* when to fly. They time their trips over the peaks to coincide with cooler air conditions that make it easier to fly and breathe.

Birds, in general, have many spectacular physical features and physiological processes that enable them to use oxygen far more efficiently than you or me. The average size of a bird's heart, for example, is 50 percent bigger than that of a mammal the same size. Their lungs have a greater capacity to make efficient use of the oxygen in their blood. Indeed, even lowland sparrows can fly in a wind tunnel while breathing air so low in oxygen, it would render a mouse comatose.

The joy of discovering such wondrous things crashes against the shoreline of widespread exploitation. How can we cage geese while at the same time registering such exquisite capacities for flight?

In such moments, I try to remind myself: *I reject fatalism.* I therefore refuse to approach industrial animal agriculture as a foregone conclusion. I check the expiry dates to see what is currently happening around this issue.

Next, I encourage myself. *I am on a reparative quest,* I say. True transformations are fraught with challenges and a requirement for deep commitment. Rather than allowing myself to ruminate exclusively on the pain story of animal suffering, I expect myself to feel all that I am feeling, to honor that truth and to seek ways to be part of the many initiatives to disrupt and replace that unjust system with something better.

Demand cage-free

The most powerful movements for change are those in which people hold diverse positions. I draw strength from the knowledge that people are working across this broad collective in a vast number of different ways to push for wider systemic change.

In the years that I could not bring myself to engage with industrial animal agriculture because I was too overwhelmed by the horror of it, I was inspired by Lori Marino (the creator of the Someone Project) and how she used her academic stature to illuminate the wondrous capacities of animals who were being actively objectified. I also tried to emulate the tenacity of Temple Grandin, who channeled her empathy for animals into her decision to work to make slaughterhouses more humane.[47]

These people gave me the courage to reengage with an issue I could not bear to look at. I continue to try to invest my energy to demand full-scale rejection of industrial animal agriculture while at the same time encouraging improvements to a system I believe should not exist. I am inspired, for example, by the powerful movement of people all over the globe choosing cage-free eggs because of concern for the way hens are raised. Restaurants and food corporations are responding, with companies displaying "ground-breaking momentum" to meet their cage-free commitments, according to a May 2024 report detailing the activities of seventy-five countries across six continents.[48] Eighty-nine percent of all corporate cage-free commitments with deadlines of 2023 or earlier have been fulfilled, according to the Open Wing Alliance, a global coalition of ninety-five animal protection organizations. "With nearly every food company around the world pledging to remove cruel cages from their egg supply chains, it's clear that cage-free is the expectation—not the exception," says director Carley Betts.

The shift toward cage-free eggs is complicated *and* fraught with issues *and* it is an important global trend. Knowing all of this empowers me. Carried along by the confidence that what I am demanding is happening in places all over the world, I remind the local grocery store in California of the caged-egg ban in the state, and convince them to stop selling caged eggs. I feel more assured and forceful in this and other actions because I know that restaurant owners and food companies and individual shoppers are also making the same demands and choices, and that collectively, we're accelerating the cage-free trend.

When we challenge the starting-line fallacy, we quickly learn that raising chickens in intensive factory farms is not an intractable problem. It only began in earnest in the 1950s. Within a decade, consumers had already begun objecting to the use of battery cages. In 2015, just 6 percent of U.S. hens were raised cage-free.[49] In 2024, that number has risen to 40 percent.[50] There is still a very long way to go, not just in the United States but all over the world, *and* we can build on the much-needed momentum of what we know works.

Responding from a position of hope isn't about condoning the injustice or suffering of raising animals in an industrial system, or looking away when governments or companies fail to honor their animal welfare promises (and the lawsuits that seem to be required to hold them accountable[51]). It doesn't downplay the magnitude and complexity of animal rights issues. All of these things are urgent and true.

You may wish to check in with your feelings as you think about the enormity of industrialized animal agriculture. Making a system in which humans dominate other animals less cruel by shifting to cage-free conditions is really important, *and* it isn't anywhere near enough.[52] Part of holding onto hopefulness is remembering the stance *I am on a repara-tive quest*. That quest requires you to develop the capacity to hold two important beliefs at the same time. One is that until the broken system is fixed, a solution has not been found. Until everything is okay, nothing is okay. There is no partial justice. The other is that within the broken system, there are examples of healing that should be celebrated and amplified. Do your best to hold one view *and* the other.

More-than-human rights

Embrace the beautiful fact that we are not, and never were, at the top of a hierarchy in the first place. What we need is multispecies justice. Inspiring changes are emerging in that direction. In July 2022, the United Nations General Assembly declared that everyone on the planet has a right to a healthy environment.[53] This, along with a prolifera-tion of rights-of-nature legislation, is part of what César Rodríguez-Garavito, an international human rights and environmental law scholar at New York University, calls an "ecological turn" in more-than-human rights that is "effec-tively blurring the categorical distinction between humans and non-humans, as well as challenging the anthropocen-trism and human supremacism that has dominated fields like human rights."[54]

Robin Wall Kimmerer expresses this so beautifully as she calls out the ways we open the doors to exploitation whenever we agree to go along with the "it-ing" of nature. "If it's just stuff, we can treat it any way that we want. But if it's family, if it's beings, if they're other persons we have eco-logical compassion for them," she explains in an interview for *Orion* magazine:

Speaking with the grammar of animacy brings us all into this circle of moral consideration. Whereas when we say "it," we set those beings, those "things," as they say, outside of our circle of moral responsibility. And so, it seems both disloyal to those plants, and just deeply, ethically wrong to set other beings who care for us, who take care of us, who bring us all of these gifts, to set them outside as if they were nothing. It's dishonorable.[55]

Our lives are inextricably entangled with the lives of other species. We can no longer pretend that it is even possible, let alone desirable, to continue to maintain a line between human rights-holders and the rest of the more-than-human world.

Eating plants is a major global trend

At the same time, a movement away from eating animals is spreading quickly around the world. It is fueled by concerns about the rights and treatment of other species, as well as by the health benefits of plant-based eating, the benefits for climate change, and other environmental and ethical considerations. There is a high likelihood that you are already shifting your food preferences toward plant-based eating. I find hope in the fact that we are in the midst of a plant-based revolution that is transforming the global food industry. Africa is already leading a plant-based future, according to a headline in *Corporate Knights*.[56] The Asia-Pacific region is projected to become the biggest market for plant-based food, with significant growth in China.[57] Latin Americans are embracing plant-based foods, and demand is growing across North America.[58] More than half of Europeans have dropped their meat intake, while 40 percent plan to buy more plant-based products.[59] As of December 2024, the worldwide market for plant-based food was valued at US$11.8 billion, and was projected to more than double to $30.3 billion by 2033.[60]

No matter where you look, this shift toward green cuisine is being led by the world's largest demographic—youth. As it grows, it is deliciously diversifying, often reviving food cultures that were previously eclipsed by Western conceptions of what to eat. As journalist Shilpa Tiwari explains,

> The narrative surrounding plant-based diets often centres on Western experiences, inadvertently sidelining the rich, diverse culinary traditions of other regions... Deep cultural and spiritual threads are woven into the fabric of plant-based diets in many African nations... Parallel to this, in the United States, African American women have emerged as leading voices in the vegan movement, illustrating a profound and transformative cultural shift.[61]

What if your whole country made plant-based eating easier?

You need look no further than India to witness how long philosophical and religious traditions of nonviolence toward other species have shaped food choice. Eight in ten Indians limit meat in their diets. Forty percent of India's 1.4 billion people identify as vegetarian.[62]

So what helps drive the motivation to shift to plant-based eating into a reality in meat-loving locations? The answer lies in creating food cultures in which eating plants becomes the taken-for-granted norm.

This is the ambition behind Denmark's brilliant commitment to make plant-based eating simply what people in Denmark do. It's an idea that has the backing of all the major political parties as well as private and public partnerships. It includes training chefs in private and public kitchens; plant-based cook-ups at music festivals; new degree programs to train vegetarian chefs; strengthening plant-based cooking skills throughout the entire education system; and an eco-scheme to support farmers transitioning to plant-protein production.

It's an inspiring move, not least of all because Denmark's agricultural sector is dominated by pork and dairy. More than half of Denmark's land is used for farming, and animal agriculture contributes about a third of its carbon emissions. (Producing plant-based foods creates roughly half the emissions compared to meat production.[63])

"Plant-based foods are the future," said Jacob Jensen, Denmark's Minister for Food, Agriculture and Fisheries, when the plan was unveiled in 2023. Denmark has pledged to reduce its greenhouse gas emissions by 70 percent by 2030, and reducing meat and dairy consumption is a key part of achieving that goal. In June 2024, Denmark became the first country in the world to introduce a tax on livestock emissions, a precedent described in *The Copenhagen Post* as "historic."[64]

It is a win-win-win scenario. Studies provide evidence that even partially substituting animal proteins with plant protein foods "can increase life expectancy and decrease greenhouse gas emissions." Replacing just half of your red meat intake with plant protein can shrink the carbon footprint of your diet by 25 percent.[65] It also brings significant savings in health-care costs.

Given the rapidly increasing global demand, the Danish government believes that the necessary shift toward plant-based eating also brings massive economic opportunities. As the 2023 Action Plan for Plant-Based Foods explains, by positioning itself as a global leader, Denmark could bring in 13.5 billion kroner (US$2 billion) from plant-based food production, as well as creating 27,000 jobs.[66]

What Denmark is doing is daring to invest in transforming its entire food culture. It's the kind of visionary leadership that made this country a successful early adopter of wind energy. The decision to put plants first is groundbreaking because it counters the narrative that "advanced," "civilized" and rich countries should rely on meat consumption. It dares to challenge meat-eating as a sign of power, wealth or masculinity. Rather than focusing on reducing meat consumption, Denmark is targeting funding and policies to promote plant-based eating. "I think it matters that the Action Plan comes from a country with strong reliance on animal production as it can show to other countries with similar agriculture sectors that this is possible," says Rune-Christoffer Dragsdahl, director of the Vegetarian Society of Denmark, in an article in *Forbes*.[67]

It's hopeful to watch other places following the trend. In 2023, 1,400 mayors of U.S. cities ratified a resolution to promote a shift toward plant-based diets to address climate change, and to reduce chronic diseases and health-care costs. It's a wise move, considering food production makes up 13 percent of cities' carbon emissions every year.[68]

Make the better
option the default

You refuse to let those in power off the hook each time you demand a systemic solution rather than accepting the idea that what you do is simply a personal decision. Look at what hospitals in New York City are doing with respect to plant-based eating, for example, and demand the same thing in the schools or other institutions where you eat.

In March 2024, the eleven public hospitals in New York City jointly celebrated serving their 1.2 *millionth* plant-based meal. Plant-based meals are now the default option for patients in care. "Scientific research has shown that plant-based eating patterns are linked to significantly lower risk of cardiovascular disease, type 2 diabetes, obesity, and certain cancers. They can also be effective for weight management as well as treatment of hypertension and hyperlipidemia," announces the hospitals' press release.[69] Shifting to plant-based food has been positive in more ways than one. It reduced the hospitals' carbon emissions by 36 percent and created a cost savings of fifty-nine cents per meal. Patients liked the change too, giving plant-based meals a satisfaction score of over 90 percent.

Prepare some
plant-based comfort food

Here's an easy recipe for Frikadeller "non-meat" balls, a veggie twist on the classic Danish comfort food, as a tiny gift to make you and the Earth more comfortable.

Ingredients

4 tablespoons ground flaxseed
   or chia seeds

¼ cup water

3–4 beets

2–3 parsnips

3–4 carrots

¼ cup almond milk

1–2 vegetable bouillon cubes

1 onion

2 cloves garlic

2–3 tablespoons flour

Salt and pepper to taste

Olive oil for cooking

Preparation:

1. Whisk the ground flaxseed or chia seeds with the water. Let it sit for 10 minutes. Pour this surprisingly egg-like mixture into a large mixing bowl.
2. Shred the beets, parsnips, and carrots and add them to the bowl.
3. Heat the almond milk and dissolve the bouillon cubes into it. Pour into the mixture.
4. Finely chop the onions and garlic and add to the bowl.
5. Add the flour, salt and pepper.
6. Stir the whole mixture together.
7. Pour off any excess liquid.
8. Form the mixture into small balls (the size of Ping-Pong balls).
9. Fry in a heated pan with olive oil for 6–7 minutes, turning to cook all sides.
10. Eat!

We have been taught to see conflict.
We can learn to see coexistence

The dead tree in the backyard is shared by all the birds—hummers, hawks, robins, chickadees and many more. The cedar is home to owls and raccoons. The more I think about land *ownership*, the odder it seems. It is a uniquely human concept that has been taught and imposed, and one fraught with conflict. "Some of the most enduring and dangerous territorial disputes often involve claims of historical ownership by at least one side of a dispute," write the authors of a 2020 study published in the *Journal of Politics*.[70]

I find it very hopeful that wildlife territories are not properties to be possessed. We are not bound by biology to reproduce this notion of individual land ownership. Many wildlife territories turn out to be common spaces in which multiple species, including predators and prey, frequently coexist. They vary and shift with season, circumstance and individuality.

It is easy to imagine that prey species are doing their utmost to stay well clear of those who could feast upon them, and yet, I find examples of coexistence that broaden my black-and-white thinking (and inspire me to think about co-housing in much more open-minded ways!). My current favorite comes from Kenya, where warthogs and porcupines have been recorded occupying subterranean dens *at the same time* as their predators, spotted hyenas.

One den, at its peak, was being shared by six warthogs, two porcupines and eleven hyenas. According to Marc Dupuis-Desormeaux, a conservation biologist, the animals all used the same entrance, sometimes less than five minutes apart.[71]

Den sharing, it turns out, goes back—way back. Two hundred and fifty million years ago, an early carnivorous mammal known as *Thrinaxodon liorhinus* shared a den with its amphibian-like potential prey, *Broomistega putterilli*—an occurrence immortalized in the fossil record. Red foxes in Japan have been known to share dens with multiple types of rodents. In Italy, porcupines, red foxes and European badgers have all been spotted cohabitating.[72]

Give yourself a treat and watch "A Wild Way to Move," a time-lapse video created by Parks Canada.[73] In it you'll see all kinds of predators and prey—deer, moose, elk, bighorn sheep, wolverines, lynxes, black bears, grizzly bears, wolves, coyotes and cougars using the same wildlife bridges and underpasses to safely cross highways. These animals are not mortal "enemies" locked in a murderous lottery of instinct. The relationships between predators and prey are far more complex and nuanced than the oft-quoted idiom "eat or be eaten" suggests.

Our understanding of how animals and other living species interact with each other, and us, forms the basis of how we believe nature works. A video of a young badger and coyote playing together in a culvert in the Santa Cruz mountains of California took the internet by storm a few years back.[74] It provided further evidence to support what Indigenous communities have long realized: Badgers and coyotes socialize and sometimes collaborate to hunt. The video is a great counterpoint to the narrow view of competition we are so often given. (As one example, a recent study found that U.S. undergraduate students held a zero-sum perception, erroneously viewing the needs of humans and animals as in competition with one another.[75]) Indeed, whether we see nature as competitive or cooperative has a lot more to do with the stories we've been told about the way animals behave than it does with the animal behavior itself. And the implications of what we believe can have a profound impact on whether we treat the greater-than-human world as an adversary or a gift.

Live *with* other species

One of the reasons the stance *I am nature* is so important to hope is that it reminds us that whatever we do in our everyday lives matters to other species. Rather than seeing the world in the ways we've been taught to see—that nature is something separate from us that exists somewhere out in the wilderness—it's time to embrace what has always been true. We are intimately connected to the lives of other species.

Embracing this is even more important on our rapidly urbanizing planet. Rather than "saving nature" we need to live collectively with other species as an integrated whole. It's a movement beyond protection toward connection. Some have taken to calling it *convivial conservation.* "Inspired by decolonial, youth and Indigenous movements, convivial conservation aims to foreground social justice in conservation efforts, highlight the importance of attending to how global political and economic systems drive biodiversity destruction, and challenge the human-nature dichotomy prevalent in conservation efforts that aim to preserve an idealised 'wilderness' separated from humans," explain the authors of a 2022 research report in *Conservation and Society.*[76]

There are many hopeful examples of where this is happening. In India, 32 million people are already living inside tiger habitats that are home to 70 percent of the world's remaining tigers. The Pilibhit Tiger Reserve is bisected by roads and flanked by farmland, a school and roadside cafes in the foothills of the Himalayas. The tiger population has doubled thanks to a concerted effort of two hundred volunteers who track tigers and work with communities on how to stay safe.[77]

One of the world's largest transboundary conservation areas crosses Angola, Botswana, Namibia, Zambia and Zimbabwe. Establishing connectivity through migration corridors and working collectively to restore habitats and address poaching has had a positive impact over the past twenty-five years on the largest land animals on Earth—elephants. Through stitching habitats back together, Africa's largest savanna elephant population is stable.[78]

It's easy to recall shocking moments when I've realized something is a lot more broken than I imagined—the magnitude of the microplastics problem, the scale of species loss, a betrayal by someone I trusted and loved. I find it helpful to consciously try to remember times when I was shocked by something that is changing for the better. I am constantly finding surprising examples of change and possibility as I learn more about other species. Sometimes when I'm feeling beleaguered by issues that must change but don't, I think of baby white shark nurseries.

Chances are high that you, or others you know, are afraid of sharks. If you're lucky enough to be swimming off a beach in Southern California, in the United States, you might be hoping sharks are nowhere near. Yet white sharks, it turns out, swim close to people almost *every day* along the coast of Carpinteria, which is just south of Santa Barbara, and Del Mar, just north of San Diego. Drone footage from a two-year study published in 2023 shows juvenile white sharks actively steering *away* from paddleboarders, surfers and swimmers just as they would from an adult white shark.[79] The study is a game changer because it's the first to document how frequently sharks and people come into close proximity. Yet bites remain exceedingly rare. The people in this busy beach region are enjoying the ocean alongside great white sharks. They just didn't know they were.

Chris Lowe, a marine biologist at California State University, Long Beach, and his co-researchers will often see sharks swimming beside or underneath unwitting humans. "For years, we've been saying we really don't think sharks are as dangerous to people as people think or as they've been taught to believe," Lowe said. "And what this research shows, for the first time, is that that's true."[80]

Witnessing people developing their capacities to live in coexistence with sharks, crocodiles, tigers, elephants and more is very hopeful. It motivates me to look beyond the speciesism that taught me to value and appreciate some animals, such as dogs, and to loathe or fear others, like rodents. For the past few years I've been making a concerted effort to overcome my prejudice against rats. I encourage you to try the following practice. Actively trying to change your mind about a species you've been taught to detest turns out to be quite a positive, transformative experience on a personal level, and it's vitally important for coexistence with other species. On a rapidly urbanizing planet, finding ways to live more integrated lives with nature is essential for biodiversity to flourish.

Choose an animal you dislike
and intentionally try to love them

1. **Choose an animal that you don't like or might even be afraid of and make a conscious choice to try to love them.**

2. **Start by checking the expiry date on what you are sure you know about your chosen species.** (Digging through research into rats quickly revealed I was carrying all kinds of misperceptions. It turns out that rats are social, empathetic animals who help each other and who remember those who are nice to them.[81] They play hide and seek, and they giggle when they are tickled. Perhaps most sobering, rats didn't actually cause the bubonic plague for which they have been globally maligned for centuries. The Black Death killed 25 million people—more than a third of Europe's population—in the mid-1300s. Guess who actually caused that epidemic? People! In 2018, a team of researchers from Oslo and Ferrara revealed the humbling truth: The bubonic plague was caused by humans, along with human fleas and body lice.[82])

3. **Confront the myths that fuel your disdain.** (In the case of rats, it's a taken-for-granted truth that they are teeming with disease. Their very existence poses an ongoing harm to us; a fact underscored by the aisle in every hardware store selling kill traps, poisons and other products to exterminate rats. Yet this idea of rats spreading pestilence and filth turns out to be an

incorrect myth. Researchers at Georgetown University in Washington studied data on three thousand mammals, expecting to find that those living in urban areas carried more viruses that could be caught by humans because they live in such close contact. They were surprised to find that although rats do indeed get sick and carry disease, rats were no more likely to be the source of a new human disease than any other animal.[83] In another study published in 2022, researchers in Vienna, Austria, captured rats over a two-year period in busy tourist areas—at a cruise-ship port, in a popular city square and along a riverside walk. They tested the rats for eight types of dangerous viruses known to be harbored in rats, including hantavirus, coronavirus, hepatitis and influenza. Not a single rat carried any of the diseases.[84] A number of other studies have found similar results. Studies like these that *don't* find the presence of disease in rats are rarely published, say the scientists who conducted this study, fueling the "misconception of the reality" that urban rats are all teeming with contagion.)

4. **Create opportunities to actually feel empathy.**
(Because I sleep outside, I often encounter rats at night when I am settling into bed. After discovering muddy rat footprints on my pillowcase one morning, I bought a live trap to try to decrease the number of rats running through my sleeping area. What I soon witnessed was the compassion of rats for one another. If I caught one, I often saw others milling around the trap trying to help the captive escape. A number of times, when one rat was caught, another would voluntarily enter the

trap, even when there was no food to lure them in. Their commitment to each other moved me. I recently read research showing that rats change their behavior when it harms others.[85] Why, I wondered, have I thought it was okay not to show them the same level of consideration? Indeed, the number of research studies demonstrating the propensity of rats to extend care and to try to assist another in distress is so rich, our ongoing refusal to deny it is tantamount to what primatologist Frans de Waal called "anthropodenial"—the stubborn tendency to deny humanlike characteristics in animals, no matter how convincing the evidence.)

5. **Think about what message you are sending.** Even if you can't yet bring yourself to fall in love with an animal you once abhorred, try your best to be very clear about what you are communicating with your actions. (When I leave my garbage overflowing, I am saying to the rats, *This food doesn't matter to me, please help yourself.* Coexisting with other species is about asking different things from ourselves. It's inspiring to see initiatives like the new regulations rolled out in 2024 in Missoula, Montana, for example, that prioritize the lives of bears by requiring residents to use bear-proof garbage cans.[86])

6. **Expect yourself to change.** (Over time, I have come to see patterns in the seasons of the life of the rats in the yard. By intentionally reading more research about rats, I've learned that the brown rats, or *Rattus norvegicus*, who are the most ubiquitous city dwellers, are serious homebodies. They rarely venture more than 110 yards, or 100 meters, from their nest to acquire water and

food, which means I'm seeing the same individuals over and over again. When the fall rains start, I see rats more often, but only for a week or so, until they find safe places out of the wet. Instead of panicking that I have a rat infestation, I now see their activity as a sign that autumn has arrived. I can peacefully live with that.)

7. **Respect the right of nonhuman species to coexist.** Learn more about the rights of plants, animals and ecosystems, such as rivers, where you live. Draw inspiration to expand those rights by learning how Bolivia, New Zealand, India, Ecuador, Panama and even the city of Pittsburgh have granted rights to nature.

Celebrate your capacity to choose to love

You can learn to love something that you didn't even like. I didn't used to believe this. I only learned that it was true when I was confronted with needing to fall in love with a toy breed dog for the sake of my daughter. She was nine at the time and desperate for a dog. Eager to soothe her desire without making a long-term commitment, we decided to foster a litter of puppies from our local pet adoption center. They were Chihuahuas—five tiny sisters—a safe bet, we reckoned, as none of us cared for toy breeds. Yet, as everyone except me seemed to know would happen, it wasn't long before my daughter fell in love with a four-pound brown-and-white puppy she'd named "Cookie Dough." On the eve of her tenth birthday, my then-husband and I sat up in bed, deciding whether to do something practical or something wonderful. We chose the latter, and the next day the dog was ours for keeps.

I say *for keeps*, but that isn't really true. From the moment we gave our daughter the dog, I was consumed with regret. Chihuahuas bark incessantly. They pee on the floor. They snap. They shiver. They can live to be twenty-one. I chronicled these well-researched flaws on a regular basis. I was a lover of big dogs, and it pained me to "own" a dog so small and skittish she scurried close to the baseboards when she crossed the room.

I offered to replace Cookie Dough with a "real dog," pointing out labradoodle puppies and golden retrievers whenever they passed by. I was so awful that eventually my son, who is a few years older than his sister, looked up from his cereal one morning and said, "Mom, you gave her the dog. If it comes to a family vote, I'm siding with her."

But it wasn't until my daughter said to me in a quiet, steady voice: "I love Cookie Dough so much that if you can't love her, we must give her away. It hurts me too much that you don't"—that something inside me shifted. "It is not up to you to give up on the one you love," I told her. "It's up to me to find a way to love her."

I called my friend Nancy in Santa Barbara. "I need to fall in love with this Chihuahua," I said. "Bring her here," she said. So I tucked the dog in the car, and on the five-hour drive down the coast I was pleasantly surprised that she hopped out and peed each time I stopped for coffee. No fuss; just a quick rest stop. Nancy and her dog met us on the beach, and when she bade me to unclip the leash, I was amazed by Cookie Dough's ability to race past dogs multiple times her size. "Now we sleep with our dogs on the bed,"

Nancy said when we returned to her house. We stretched out on the duvet and Cookie Dough curled up snugly in the crook of my arm. "And now, the dogs play while we drink wine," said Nancy, uncorking a Pinot Noir. And so they did and so did we. Two days later, I drove back home with my friendly, playful, adaptable pup. Fourteen years later, Cookie Dough is well and truly *my dog.* I never knew you could make yourself fall in love, but now I know it's true. We can with other species. We can choose to love.

You are not living in a *dog-eat-dog world*

I encourage you to look anew at whatever assumptions you might be carrying about other species, and about the way the world works. The ruthless self-interest expressed in the idiom that we live in a *dog-eat-dog world,* for example, misrepresents most people—and most dogs! Dogs, and the wolves from which they evolved, are highly cooperative species.[87] And studies of people from a wide range of cultures consistently show that "even in consequential real-world situations, humans are not narrowly self-interested but substantially generous."[88] Small acts of kindness are frequent and universal, according to an international study in 2023.[89] In an analysis of video recordings of people around the world doing daily activities, the researchers found that on average every two minutes, someone asked another for assistance and help was almost always given.

This impulse to help starts early, according to Felix Warneken, director of the Social Minds Lab at the University of Michigan. For almost two decades he's been watching toddlers. His research reveals that very young children have

a spontaneous tendency to care about others that is biologically based. They help early on, without being asked, when
they aren't being observed by their parents, and without
being offered any type of reward.[90]

Experience awe through
the goodness of others

When researchers at the University of California, Berkeley,
asked people in twenty-six countries what made them feel
a sense of awe, the responses were about witnessing acts
of moral beauty: things like acts of generosity, strength,
or courage to overcome obstacles. People experienced awe
while watching a bystander interrupting a crime, someone
overcoming incredible adversity to achieve great things, or a
crowd standing in solidarity to improve the lives of others.[91]
Hope rises when we witness our collective capacity for good.

I'm currently working with Ginny Broadhurst, director
of the Salish Sea Institute, and environmental leaders across
the Salish Sea region on a hope network project. Collectively, these leaders have chosen to signpost positive changes
occurring in the region, like the remarkable ecosystems that
have emerged in response to the removal of the Elwha River
dam, and the astonishing recovery of kelp associated with
the reintroduction of sea otters. They are refusing to disempower themselves, choosing instead to find their collective
courage to demand more badly needed changes by focusing
on what they know works.

Dam removal is a case in point. In 2023 alone, eighty
dams were demolished across the United States. Barry
McCovey, director of the Yurok Tribal Fisheries Program,
remembers when dam removal seemed an impossible

dream—how tribal members were laughed out of the room when they proposed removing the Klamath River dams in California. Then in 2002, the deaths of seventy thousand endangered salmon sparked a tribal-led movement of Native and non-Native voices, policymakers, lawyers, scientists and others, all united under the rallying cry of "Undam the Klamath."[92]

On October 5, 2024, the largest dam-removal project in history was completed: 420 miles (676 km) of the Klamath River and its tributaries are once again free flowing, thanks to the removal of four hydroelectric dams.[93] Tearing down dams is a powerful form of ecological restoration. "There's nothing better we could do for fish and for people in Northern California and throughout the world than to remove these dams," according to Robert Lusardi, a freshwater ecologist. Indeed, each time a dam is removed or an estuary reopened along the west coast of North America, the story is the same: the fish return immediately, often on the very first day.

To be part of a reparative quest of transformation unfolding on such a mammoth scale is life-defining. Speaking to the BBC in November 2024, Brook Thompson, a Yurok and Karuk PhD student who chose a civil engineering degree in order to pursue the restoration work she is currently doing in the Klamath basin, says, "Salmon returning to the upper Klamath River the same season as the dam removal has filled me with gratification and joy that I usually reserve for weddings and births . . . the salmon returning home brought me to tears of happiness and relief. We did it. I am filled with newfound energy and hope to keep up my work in restoration and supporting the rights of tribes and fish."[94]

Host a gathering and ask,
What has changed for the better?

Focusing on successes is empowering. It is respectful of the competence that exists within other species and within ourselves. Do not allow yourself to conform to a worldview that makes you deny resilience because you are afraid successes will be used as excuses for no further action.

The more knowledge you have about what has changed for the better, the more empowered you feel to make change and the more informed you are about what actually works. Crowdsource this critical element of transformation by hosting a potluck, and ask each of your friends to share a story of something that has changed for the better that they never imagined would happen. (The examples can be at any scale and from any context: e.g., a societal shift, a human rights victory, a species recovery, a recovering environment, a personal triumph, a relationship dilemma solved, a global-scale event such as the successful repair of the ozone hole, and more.)

No more "pain" stories

It's perhaps understandable that because we love the Earth and are concerned about its destruction, we often talk about animals and plants as helpless victims. It may be that we believe that if we focus on the horror of deforestation and urban sprawl and so much more, we'll try harder to help other species.

Unfortunately, it's a flawed theory of change. I've learned a lot about the importance of rejecting these kinds of "pain" stories from the work of Eve Tuck.[95] Eve is Unangax̂ and is an enrolled member of the Aleut Community of Saint Paul Island, Alaska. She is also a professor of Indigenous studies at New York University.

There is no question that injustice and harm must be called out. Yet by focusing strictly on the damage that communities face, we create pathologizing narratives of pain and brokenness that define communities entirely by oppression. When we only create space to foreground the hardship, we silence the resilience that also lives within each remarkable individual and group. I fear we are doing the same thing to other animals and plants.

Ever since Eve opened my eyes to the problem of narratives centered on damage, I've noticed what a common default they are. Photos from the 2019–20 bushfires that devastated Australia, for example, depict charred trees and wounded animals. These are terrible truths and they deserve front-page coverage. In testament to the number of deaths the fires inflicted, the grief they left in their wake and the funereal silence of burnt forests where no leaves stir, the season was given the name "Black Summer."

Yet the global headlines heralding the deaths of between one and three *billion* animals were pain stories run amok. Those huge numbers seem to presuppose that all animals can do is stand in the path of a fire and burn to death. I want to be clear that the scale of the fires and the loss of life were horrendous. "Black Summer was the first time that bushfires burnt so vastly and for so long that they were neither a confined 'event' and nor were they just 'Australian,'" writes Tom Griffiths, professor of history at the Australian National University. "They became a planetary phenomenon: Smoke from these fires encircled the globe."[96]

I do, however, also want to point out that the response of animals to wildfires is complicated and full of agency. Animals are not passive victims—a fact that Chris Jolly, a scientist at Charles Sturt University, and his colleagues underscore in a global synthesis paper from 2022. Their research, as he explains in an article in *The Conversation*, "suggests that, on average, the vast majority of animals (more than 90%) actually survive the immediate passage of a typical fire."[97]

Knowing this does not diminish the wrongness and horrific impact of climate change, drought and bushfires. Instead, it puts us in a more informed position to respond to and assist animals and the forests upon which they depend in the aftermath of such devastation. We almost never hear about what happens *after* the worst. What we didn't see in the news, and what we have a right to see, is that recovery is also occurring. By 2023, more than a thousand motion-activated cameras had gathered more than seven million photos, many showing vulnerable species returning to the devastated region. "We're finding that native species are more resilient than we thought," says Conservation International wildlife scientist Jorge Ahumada in a 2023 interview.[98]

In 2024, scientists reported more koalas in eastern Australia than they had previously thought were there. The encouraging census came from passively *listening* for the sound of koalas in a multiyear study across 224 forested sites. We need to know about these recoveries, and about the capacity for resilience that exists within plants as well.

"Eucalypts are not interested in dying. They're survivors," writes Gregory Moore, a biologist at the University of Melbourne, pointing to the impressive capacities of these trees to survive brush fires.[99] Tucked beneath their bark lie thousands of dormant buds that spring to life as leaves directly from the tree's trunk, enabling them to photosynthesize and heal even when all of their branches have burned.

Biodiversity can rebound after bushfires. I felt my heart leap when I read this conclusion of a 2023 study of changes in species diversity following the Black Summer fires in New South Wales, Australia.[100] Yet the specificity of recovery is also crucial to recognize. Species diversity significantly decreased in forests that endured the most severe fires.

We need to know all of this: Extraordinary resilience exists, and we need to recognize the specificity of each forest and each fire. Each forest is its own unique community of trees. Each fire has its own particular range, frequency and intensity. Specificity and context matter to the success of the solutions we enact.

Hone your ability to hold the good, the bad and the nuance in between. To do so is far more empowering than being rendered hopeless by sweeping apocalyptic depictions. Keep challenging yourself to honor resilience while calling out the vital need to tackle climate change and other issues causing devastating consequences. Holding complexity, awareness of injustice and resilience together at the same time is critical to the practice of evidence-based hope.

IN THE OCEAN, scientists have identified certain species of coral and coral reef fish that are better able to resist ocean temperature rise and other negative effects of climate change. When we know this, and we combine it with the knowledge that marine protected areas increase the ability of ecosystems to recover from coral bleaching, we feel more hopeful about and committed to our climate change actions. We're more likely to understand why it is imperative to hold governments accountable to their worldwide commitments to protect or restore 30 percent of land and water by 2030.

The other day I came upon a menu of what dinner in Vancouver's Burrard Inlet looked like five hundred years ago. Sweet, buttery salmon; clams tended in coastal gardens; shimmering schools of herring... this vision of a delicious, abundant ecosystem of food was assembled by the Tsleil-Waututh Nation and a team of researchers.

What is so impactful about this work is that it fully recognizes the profound destruction that exists in this big-city industrial shipping zone. Yet it also views the inlet as an empowered entity, actively intent on living. Instead of delivering a funeral eulogy, the Tsleil-Waututh Nation asks, *What is this land capable of?*

Living into this empowering position is profoundly hopeful. "They tried really hard to demolish us and remove us from our land and erase our Indigenous way of life," Tsleil-Waututh councilor Charlene Aleck says in an interview for *The Narwhal*. "We're still here. And the same with the land," she says—and the inlet is "trying to thrive."[101]

While there is endless work still to do, years of ecological restoration—replanting eelgrass meadows, restoring salmon habitat and removing creosote-treated pilings—is making a difference. Amid an industrial shipping zone, herring have returned—something many of the people working to make this happen had never experienced in their lifetime. The herring, in turn, bring sea lions, seals, eagles and herons. As single species recover, whole ecosystems rebound.

Ask, *What is this place capable of?* Shout out the resilience that exists

Choose a specific place or species that matters to you.

Ask yourself: *What is this place or species capable of becoming?* (No matter what condition it is currently experiencing, try to look upon it not as less than it once was, but as what it is actively trying to become.)

Then ask yourself: *What am I capable of offering to help that vision become a reality?*

Commit yourself to supporting that transformation.

Keep a list of other recovering places and species to give yourself ongoing courage and inspiration.

Shift from damage to desire

Eve Tuck has an inspiring concept of desire-based frameworks for understanding marginalized communities. Desire-based frameworks encourage us to fully acknowledge the pain of the past, while at the same time recognizing the wisdom and resilience that also exist within communities. I'd been thinking about it for a while, but it wasn't until I heard the song "River" on the wonderful album *Watin,* by Toronto-based artist Aysanabee, that I could feel what Eve was describing. I am moved by the many ways artists enable me to comprehend things I sometimes find difficult to grasp. "River" is based on the story of when Aysanabee's grandfather and grandmother met while at McIntosh Residential

School: "I really kind of wanted to showcase, yeah, here's a really terrible thing that's happening to a person. But here's this person kind of rising up, rising to the occasion in this moment as a child," Aysanabee said in a 2022 CBC Radio interview.[102] It's an incredible song situated in intergenerational trauma *and* love: It respects the grief while celebrating the promise of a new life together.

Keep a wide-open mind about statistics of doom

Statistics about rapid rates of species extinctions drive panic and feelings of fatalistic doom. You may have heard that 150 species go extinct every day, or that 30 percent of all animals and plants could be extinct within a hundred years. And yet, no one actually knows how many species there are on Earth.[103] Most species have never been identified. Which means all the statistics you hear about how quickly species are disappearing from the planet are based on computer modeling and extrapolations at best. "Documented losses are tiny by comparison," writes Fred Pearce in an article published by the Yale School of Environment.[104] Even prominent scientists cite numbers that vary dramatically from one another when estimating the rate at which species are going extinct.

That's not to say extinctions and other dire threats to species are not urgent and real. They absolutely are. "Whatever the drawbacks of such extrapolations, it is clear that a huge number of species are under threat from lost habitats, climate change, and other human intrusions," explains Fred.

Knowing this makes news of species recoveries even more uplifting. The global conservation status of birds appears to be improving, according to a 2024 update to the IUCN's Red

List of Threatened Species. Forty-six fewer species of birds are considered to be threatened with extinction than was the case two years earlier. This positive change in direction is due in part to a rise in cooperative, ecosystem-scaled initiatives. In 2024, for example, thirty countries committed to restore habitat that hundreds of migratory bird species use in their journey across the Central Asian Flyway.[105]

Remind yourself not to become disempowered by pessimistic projections. This is a difficult thing to talk about, and yet it is really necessary to do so. Environmental destruction is widespread, terrible and wrong. It brings with it untold hardships for all kinds of species, and many people. Yet at the same time, it's essential to appreciate just how resilient and generous life on Earth is. I find it almost too moving to express my gratitude that despite the horrors we have ravaged on the planet, I still receive the remarkable gift of waking each morning to a sky full of air.

For me, examples of resilience—and the complex relationships that enable life to exist in such impossibly compromised circumstances—are powerful motivators. They drive me to redouble my efforts to protect and restore vital habitats. We must stop cutting down old-growth forests. Full stop. And, in the way-too-many places forests have been cleared, we must call upon traditional and Indigenous knowledge and findings from Western science detailing how quickly forests can recover to help them heal. We cannot give up because we falsely believe that they are beyond hope.

You may find yourself struggling when you read examples like these. Recognizing beauty or positive developments in the midst of injustice, crises and losses can trigger fears that you are being duped or might somehow be complicit in downplaying these terrible tragedies. Once again you may grapple with the feeling that by acknowledging resilience, you will provide those in power with even more excuses to avoid their responsibilities to make vitally needed changes. These are the moments where I find Eve Tuck's concept of pain stories so powerful. She reminds me that no matter how well intentioned it is, focusing only on what's broken as a way to make change disrespects the individuals or communities involved. The most powerful way to achieve justice and fuel empowering hope is to tell the truth about wrongdoing *and* celebrate and support resilience. My responsibility is to seek out and tell fuller and more honest stories by actively embracing the complexity and ambiguity that actually exists.

But what do you do if the story *isn't* hopeful?

This is the question that emerged at a hope gathering of environmental leaders in the Salish Sea in 2024. It was posed by a documentary filmmaker who was putting together a story about Southern Resident killer whales. This distinct population of killer whales spend most of their time in the Salish Sea. They are critically endangered. They face an accelerating risk of extinction, according to a study released in 2024.[106] Their entire population numbers just seventy-four individuals.

When the filmmaker first put the question to the circle, there were many sad nods of agreement. Some stories are just hopeless. And then amid the list of overwhelming threats the whales face, the conversation slowly began to turn. "When a killer whale dies we have death ceremonies," said Stz'uminus First Nation member Ray Harris. "We've had three naming ceremonies in the last four years to welcome the births of killer whale calves. The last one we named Hope Child." His choice to highlight the whales' resilience, as well as their plight, spurred others to share the good along with the bad. Someone mentioned how in years past people stopped the live capture of killer whales for marine shows, after a third of the Southern Resident population had been taken. How interim sanctuaries and speed-restriction zones have been implemented as part of a 2024 management strategy to protect the whales. Someone else talked about how British Columbia's ferry system is shifting their fleet toward quieter electric ferries. Another mentioned how the restoration of salmon habitat and other ecological repairs are helping to give the whole ecosystem in which these whales live a better chance for recovery. Others spoke about how increasing respect for Indigenous knowledge is enhancing the recovery of the Salish Sea, and how stronger environmental laws and regulations are tackling harmful chemicals.

The conversation shifted from a dead-end tale of hopelessness to a messy story of "ands." How the ban on open net-pen salmon farming in B.C. (set to take effect in 2029) will decrease the risk of disease and the competition for food and spawning areas for the wild salmon the whales eat, and how in recent years, the whales have been spotted feeding as far south as Monterey, California, in winter and early spring, expanding or reclaiming (no one is sure which) their access to fish.[107]

Including the "ands" enabled the filmmaker to see a much more nuanced and hopeful story, which both acknowledged and then moved beyond the pain. The years of overwhelming pressures on the whales from boat traffic, pollutants and loss of their primary food source (Chinook salmon) are real. And so too are the efforts people are making to counter those wrongs. And so too is the perseverance of the highly intelligent, close-knit members of the Southern Resident killer whale population. "The risk of extinction is not a give-up-hope story," said one of the people in the circle. "Elephant seals, whooping cranes, bald eagles, all kinds of animals have made remarkable recoveries from dire straits. This is a story of never giving up."

The more proficient you become at complicating stories, at including the ambiguity and contradictions and triumphs and losses, the more adept you become at seeing the evidence-based hope that exists within the problems. This, in turn, increases your empowerment to build on that hope. The best way I know to do this is to make sure that every story I tell always includes the "ands."

• PRACTICE •

Tell stories of "ands"

1. Choose an issue that you feel is hopeless.

2. List the real problems that exist.

3. Invite others to help you complicate the story by adding as many positive "ands" as they can think of.

4. Pause to feel the breadth and complexity of what has emerged.

5. Tell the story.

6. Check in with how you feel.

Be curious about other species

Paying attention to what other species do opens up new thinking about effective ways to help ecosystems heal. Recent research reveals that fish turn out to be surprisingly chatty animals. Of the 1,252 fish species studied, more than a thousand of them rely on sounds to attract mates, protect territories, find their way and so much more. Some species of fish sing together at dawn and dusk, just like the choruses of birds who greet the sun each morning.[108] Indeed, the fossil record reveals a delightful surprise: singing began in fish and was later perfected by birds.

At this very moment, tiny larval fish smaller than a fingernail clipping are listening for the clicks, hoots, pops and grunts of fish and invertebrates living together on coral reefs. They're listening their way toward the sound of healthy, thriving reefs. So great is the draw of these promising sounds of food and community that scientists are actually using recordings of healthy coral reefs to attract invertebrates and fish to reefs that have grown quiet due to coral bleaching.[109] Not only fish, but corals too, set out in search of new reefs when they are babies. Researchers with the Woods Hole Oceanographic Institution discovered that mimicking the sounds of healthy reefs caused coral larvae to recolonize and begin to heal damaged reefs at rates seven times greater than would have occurred without the additional sounds.[110]

Scientists studying an Indonesian reef that had been devastated by blast fishing (an illegal fishing technique that uses explosives to blow up underwater habitat in order to strategically target schools of fish) were thrilled to discover that efforts to restore the corals were so successful, the reef sounded much like healthy reefs that had not been damaged. Three years into restoration, they were listening to the hopeful sounds of an ecosystem coming back to life.[111]

Ecosystems are driven to heal, often faster than you'd expect

What each of these examples demonstrates is the astonishing capacity of ecosystems to heal. Tropical forests burned and cleared for farming and ranching in West Africa and South and Central America, for example, turn out not to be lost forever. If left to themselves, and protected from further harm, they can regain their lush biodiversity, creating rich havens that also suck carbon from the atmosphere. And they do it surprisingly quickly. Scientists writing in a 2021 issue of *Science* were surprised to see how fast the soils recovered. Nitrogen, carbon and soil density reached 90 percent of the levels in untouched forests after one to nine years. The size of tree leaves, the density of their wood and other functional qualities of forest plants attained milestones sooner than predicted, reaching old-growth conditions in just three to twenty-seven years.[112]

It is also true that it took between twenty-seven and a hundred and nineteen years for the total mass and size of trees to approach their former conditions. And yet even that feels remarkably quick given the complexity of tropical forests. "These regrowing forests cover vast areas, and can contribute to local and global targets for ecosystem restoration," said Lourens Poorter, an ecologist at Wageningen University in the Netherlands.[113]

What you choose to plant
can have a very big impact

It's easy to assume that what we plant in one small garden, whether it's a balcony planter or the yard of a house, doesn't make much difference in the grand scale of things. Yet when it comes to helping the recovery of monarch butterflies, it turns out what you plant creates significant habitat. In many metropolitan areas of North America, single-family homes make up the second-largest source of potential plantable space for milkweed, the host plants for monarch larvae. Community scientists working with the Field Museum's monarch project in Chicago determined that even a single milkweed plant can provide a safe harbor for butterfly eggs.[114]

One of the most inspiring personal habitat restoration projects I know of began with just two people—Brazilian photographer Sebastião Salgado and his wife, Lélia. The project started on the family farm he inherited in Vale do Rio Doce, in the state of Minas Gerais. In interviews, Sebastião describes how he and the land healed together—he from witnessing horrific events as a photographer documenting the 1994 Rwandan genocide, and the land from being deforested.[115] It was Lélia who asked him, "Sebastião, why do you not replant the forest?" And so they did.

In 1998, using seed funds generated by his fame as a photographer, they set up Instituto Terra, a nonprofit organization focused on environmental restoration in the Vale do Rio Doce. Over time they created a nursery, a laboratory for seeds and a training center that enrolled twenty environmental technicians per year. They are the first people to plant a rainforest on a large scale in Brazil.

More than three million trees have now been planted. In an interview for *National Geographic* in 2022, Sebastião says: "Within our farm, we've brought back a huge amount of biodiversity. We have practically all the insects of the region, 173 different bird species, mammals—we've even seen jaguars. Now these pockets of biodiversity are starting to radiate out to other areas."[116]

Today, the project spreads far beyond the farm. Progress is underway to rehabilitate the water basin of the entire Vale do Rio Doce—an area the size of Portugal. As the springs and streams come back to life, they return drinking water to lands crushed by erosion and the hooves of cattle. "To rehabilitate a source of water, you must plant on average 500 trees in one hectare of land," he explains. "So far, we've planted about 2,100 such small forests, and we've just received financing to plant 4,200 more in 3,000 different farms."

Sebastião turned eighty years old in 2024. I wonder what it feels like to have played in a lush forest as a child, witnessed its destruction and then orchestrated its rebirth all within one's lifetime. It's such a poignant, complicated and hopeful combination of tragic violence and reparative quest. Speaking to a German reporter that year, Sebastião says: "Life returned—insects, birds, the first mammals—and as nature began to thrive on the farm, my will to live returned, too."[117]

Create a meadow. Grow a forest.

Interested in planting a monarch butterfly habitat? Here's the simple recipe the team from Chicago's Monarch Community Science Project suggests:[118]

- Plant as many milkweed plants as you have room for. (Research the best species for the area where you live.)

- Surround them with a diversity of pollinator-friendly flowering plants. (These provide the monarchs with nectar throughout the season, which allows them to stick around the area longer and lay more eggs.)

Upscale to a rainforest? Here are Sebastião's tips:

- It's important to build a forest in sequence. First plant the "pioneer" species, which will create the conditions for the forest to grow.

- Next plant the secondary trees.

- Two decades after you begin, plant the "climax trees." (Some of these more durable species will disappear, but those that thrive may live for a thousand years.)

The greening of cities

In recent years, much work has been done to welcome plants back into cities. I am grateful to researchers who undertake the mammoth task of trying to quantify global trends so we can better understand if we are moving in the right direction. A 2023 study assessed 1,028 cities to understand the state of urban green spaces—like parks, gardens, woodlands and city squares—worldwide.[119] It showed that since 2011, these spaces, and people's access to them, have been increasing. This is a hopeful trend for health, biodiversity, climate justice and more. At the same time, this study and others underscore the inequity at play: cities in the Global North are gaining green spaces at a greater rate than those in the Global South. Both situations are true, and so is another finding: There is a global trend toward improving the access of people living in underserved communities to urban green space.

Maintain your hopeful stance by recognizing the complexity of the "ands" and seeing the overall trajectory as part of our reparative quest of transformation. Much work remains to be done on increasing urban green spaces and making them more equitable, *and* overall, this effort (as of May 2024) is on a positive trajectory.

So how might you use this new knowledge to help amplify this much-needed trend? I often share a quick message with mayors of cities that are actively working to increase urban green space. I thank them for what they have accomplished, and I share the current research to provide them with evidence that they are part of an important global trend. We know from psychological research that pride and sense of purpose and belonging are all important to keep us working toward difficult things. I share current information as a tailwind to push forward their efforts. I also share what I'm learning with environmental organizations and activist groups. It's hard for any of us to keep up to date on every issue.

When you find meaningful research on a current positive trend, take a moment to share it directly with someone who you know cares about the issue. This strengthens the cause and your relationship with that person and their community. It's another valuable way to swim in the waters you wish to inhabit. Rather than sending a mass communication, you are letting that person know that you specifically appreciate them and their efforts. As Charles Vogl, author of *The Art of Community*, describes so beautifully, it is a much different thing to see a post about a gathering than it is to be personally invited. The most valuable invitations are those where the person knows that you thought about them, and that if they didn't come, they would be missed.

A coexistence code of conduct

I'm a big fan of the Stockholm Resilience Centre. They had a post on their website in 2024 encouraging reflection on what it means to be a "good" urban animal. Is there a coexistence code of conduct that we need to learn? How can the rules of the game, set by humans, be meaningfully communicated to wild animals living in the city?

Here's a list of etiquette rules you might follow to coexist peaceably with other urban animals. Feel free to add to it or modify the guidelines to suit the place where you live.

Commit to interspecies etiquette in the city

1. **Be curious.** When you come across a deer crossing a street or an owl perched on a fence, carefully observe their body language. Try to deduce what that individual is engaged in doing.

2. **Be respectful.** Treat whatever nonhuman animal you meet with respect. Give them space to maneuver. Restrain your own movements so they feel seen and taken seriously.

3. **Be patient.** Other species may move at a different pace or be paying attention to different sensory cues than you are. Be quiet. Don't overwhelm them with quick movements.

4. **Be nonthreatening.** Be aware that not only you, but perhaps your dog or an oncoming car or sirens, are all potentially frightening. Assess the wider situation and intervene, if necessary, to create a safe environment.

5. **Be nonselfish.** Welcoming other species to graze or drink or rest or find refuge in the green spaces that surround where you live or work or play is a necessary and fair way of recognizing that cities are habitats in which many different species live.

CONTINUED ▶

6. **Be trustworthy.** Each time we treat another animal with respect and care, we are investing in a positive relationship. It is a wondrous experience when a wild species responds with curiosity rather than fear. Nurture the actions that grow interspecies trust.

7. **Create more wildlife habitat.** Look at wherever you live from the perspective of other species. Provide water. Make your windows bird safe. Stop using pesticides. Cover swimming pools to prevent accidental drownings. Keep cats inside. Leave dead flower heads, plant stalks and leaves where they fall to create habitat for insects. I am grateful to the neighbors who plant pollinator gardens, the cities that invest in urban trees and green roofs, the individual people across the United States who have transferred 61 million acres into land trusts.

8. **Be bold.** Treat urban greening as the social justice issue that it is. You and other species have a right to live in healthy natural environments. Expect and demand that from the communities, cities and countries in which you and many millions of other species live.

Savor your wild
collective effervescence

Despite centuries of narratives, social structures and norms that have worked to disassociate humans from nature, our wild selves feel the deep connections we share with other species. That is a very beautiful and hopeful thing. We are, in fact, quite good at picking up on the emotions of animals, even species that express their feelings in very different ways from you and me.

In an intriguing experiment, people were asked to listen to the hoots, squeals and chirps of nine different species, from hourglass tree frogs to American alligators. Some of the sounds were made by animals when they were excited, perhaps reacting to a threat or competing for a mate, while others reflected a relaxed state. Regardless of whether the listeners spoke English, German or Mandarin, they were good at interpreting the emotional state of birds, mammals, amphibians and reptiles.[120] A more recent study of participants selected from forty-eight countries listening to different types of mammals found similar results, with an encouraging caveat: People who regularly work with animals performed best at understanding the emotional content of animals' vocalizations.[121]

We are in sync with other species. This inspiring discovery reminds me of a magical thing I learned from Nicole Furlonge. MRI experiments, she told me, reveal that simply by listening to stories told by another person our brains fall into rhythm, mirroring the wave pattern of their thoughts as the tale unfolds. More miraculous still, our minds begin to anticipate where the story will go. This happens even when the person we are listening to is speaking a language we do not speak.[122]

Such extraordinary responsiveness between one life and another, regardless of species, makes me think of the beautiful manner in which adults and babies sync up during play. "Their brains influence each other in dynamic ways," says Elise Piazza, lead author of a 2020 study conducted at the Princeton Baby Lab. As her coauthor Casey Lew-Williams explains, "The infant brain was often 'leading' the adult brain by a few seconds, suggesting that babies do not just passively receive input but may guide adults toward the next thing they're going to focus on: which toy to pick up, which words to say."[123]

This capacity for our minds to connect can surge through a music hall or a sports arena. Sociologists have a term for a group emotion that is shared across fans or concertgoers participating in a stirring performance. They call it *collective effervescence.*

It feels spiritual: a connection that soars across hearts and souls. I think of the chimps in Tanzania, leaping with excitement in the spray of a magnificent waterfall. Later, when the exuberance subsides, they sit quietly on rocks where the water pools, gazing in reverence at the spectacle before them. "I can't help feeling that this waterfall display or dance is perhaps triggered by feelings of awe, wonder that we feel," says primatologist Jane Goodall. "The chimpanzee's brain is so like ours, they have emotions that are clearly similar or the same as those that we call happiness, sadness, fear, despair and so forth; incredible intellectual abilities that we used to think unique to us, so why wouldn't they also have feelings of some kind of spirituality?"[124]

I think of whale watchers all over the planet, gasping in delight at the sight of a blow; urban birders, craning to see a rare species just a whisker away from view; gardeners, hands deep in the soil, planting their faith in a harvest yet to come. It makes my heart sing to witness our collective effervescence with nature.

IT'S ALREADY a blue-sky morning and it's only 5:00 a.m. Bird calls dart through the trees like a hopscotch symphony. I'm sitting on the back deck, waiting to speak via Zoom with a group of students in Turkey nine time zones away. As I contemplate whether to sneak back into the house to make a cup of tea and risk waking others, a gust of wind threatens to blow my laptop off the table. Wind rushes through the top of the Garry oak trees that dwarf my neighbor's house, sashaying their upper branches in a dancing flurry of spring-green leaves.

This wind feels so genuinely alive, I find myself pondering why I accept it being categorized as a lifeless element. Surely it deserves more credit than that. Wind scatters the acorns that grow into these massive trees. Wind carves the mountains and powers the strongest ocean currents. Wind blows through the universe, seeding the formation of galaxies.[125]

It makes my heart happy, therefore, to discover that within the written and oral cultures of Damara / ≠Nūkhoen peoples in Namibia, winds *are* alive.[126] Winds are loving and caring. They have distinct personalities, genders and intentions. They are a couple. They make joint decisions. Only if she, the westerly sea wind, and he, the eastern wind, agree to come together will they bring the clouds and ultimately, the rain.[127]

Damara / ≠Nūkhoen cultures consider the perspectives not only of other people but also of other animals, such as elephants. They extend their empathy to a wide range of beings, including the wind. In this multispecies worldview, all of these entities possess a diversity of feelings, choices and desires.[128]

I aspire to live in such vibrant and openhearted ways. To see beyond categories of living/nonliving or sentient/unfeeling. I am no longer content to be bound by such stingy views. Wild contagious hope rests in the celebration of truth telling, complexity, connection, abundance, repair and restoration.

HERE IS A FINAL practice of hope that you can do anywhere, anytime. It is an intimate reminder of the infinite elegance and generosity of life on Earth. Draw in a deep breath and contemplate all of the intricacies that make this simple, powerful action possible. Wonder at how your body practiced breathing amniotic fluid even before you knew air existed, while you were still inside the womb. Appreciate the way your first breath at birth unleashed a cascade of miraculous changes in your heart, kidneys, endocrine and digestive systems, lungs and brain that have made all the other moments in your life possible. Like your arms and legs, your lungs grew throughout your childhood. Eighty percent of the tiny air sacs that transfer oxygen to your blood developed while you were still a kid.

Choose eight trees in your neighborhood and create a simple ritual to thank them for their thoughtful care. Eight trees—that is roughly how many it takes to create the air you breathe.[129] If you find yourself by the seashore, honor the kelp and seaweed. Stand by the shore and shout words of gratitude to the swirling fronds of green that dance upon the surface. Half of all the oxygen you breathe comes from ocean plants.

Over the course of your lifetime, these brilliant green beings will help you to take about 600 million breaths. Breaths you will share with other people, other species, the whole wide world. Every breath is a new opportunity. It is the ultimate act of co-creation. The Earth gifts you with air and what you breathe back into the world shapes what life will become. Your breath is always with you. Your breath is the embodiment of hope.

Special Appendix for Educators

Teach in ways that validate eco-emotions

How often do you find yourself navigating your own journey with climate emotions, while trying to offer sensitive support to someone else who is struggling with existential dread or fear about the future?

As you saw throughout this book, no matter where or when you are talking about or working with climate issues, you are engaged in an emotional experience. You do not have a choice about whether to *make* the subject emotional. It *is* already emotional. I understand that this is a difficult idea to accept. Teachers and professors, for example, often say to me, "I know how to teach climate change, but I am not a trained therapist. I don't know how to teach emotions. I am not comfortable bringing emotions into the classroom." Yet avoiding discussing climate change because it is distressing is a disservice to students who are already navigating the real-world effects of the climate crisis. Students, teachers and professors arrive to class full of feelings about climate change and all kinds of other issues.

Pretending that emotions are only there if we allow them to be and choosing to teach climate change as if it were simply a series of facts is emotionally invalidating. Not only is psychological distress detrimental to learning, it risks being another way we may be unintentionally complicit in betraying students' trust. Schools and other educational institutions promise to be safe spaces for learning. Yet each time we teach devastating content, for instance about species extinctions or food insecurity or forever chemical pollutants, without acknowledging and supporting the emotions of students and teachers, we cause harm. Silencing is a form of interpersonal injustice.

So how do you teach or engage in climate justice conversations in ways that validate rather than dismiss or ignore emotions? In their 2024 research article on "Strategies for Delivering Trauma-Informed Climate Change Instruction," Nikki Hurless and Na Young Kong say:

> Discussing difficult topics in sensitive ways with intentionality can help children understand the reality of climate change without causing undue harm... Through a trauma-informed approach, teachers can apply action-focused and strengths-based strategies to help students understand and address climate change rather than avoiding it. Building students' confidence and competence regarding difficult, and very real, global issues is key for students to develop resilience, critical thinking, [and] empathy as adults.[1]

Whether you're speaking to a young child or an adult, we all need safe spaces where we can acknowledge and share and be accepted for how we feel before we can move on to ways we might want to act upon those feelings.

Create a safe space
for sharing eco-emotions

Begin by taking stock

1. Remind yourself that everything to do with climate change is emotional.

2. Check in with your own feelings. Are you feeling defensive, guarded, fearful, anxious, numb, peaceful, etc.? Are you feeling calm, emotionally able and equipped to offer support?

3. Make an intentional decision to let go of any expectations that you can help. Remember, this is about supporting whatever the other person is experiencing, not about fixing a problem.

4. Take a deep calming breath. Ground yourself. Quiet your mind. Feel the support of the Earth or the floor beneath your feet. If you feel it would be helpful, invite the other person to do this with you too.

Hold a safe space for whatever feelings emerge

1. Listen generously. Acknowledge and appreciate what is being shared both verbally and nonverbally.

2. Remind whoever you are speaking with that they are not broken. Their feelings are a reflection of the depth of their love and concern. Reassure them that anxiety and worry are normal and understandable reactions.

3. Focus on feelings. This is not the time to engage in a debate about the veracity of climate change.

4. Be genuinely curious and compassionate. Asking and listening are more important than seeking answers.

5. Open up spaciousness for healing, coping and resilience. There is no rush to resolve eco-anxiety or other feelings.

6. Find your strength to encounter other people's anxiety, darkest fears and other difficult feelings. (One of the things young people often express is that they want adults to be secure and committed enough to be able to bear their feelings and not quickly try to pass them off to someone else.)

If someone asks for advice on what actions to take, source solutions together

No one should be left to face climate emotions alone. When specific issues that are concerning someone surface, let them know you are willing to think about the issues together, even if you don't have any ready-made solutions. Offer to go looking for more information together. Use your familiarity with solutions journalism sources to help find meaningful content. Contain your desire to race into cheerleading action. Try to help the person find the right balance for them between rest, recovery, keeping up to date with solutions sources, and action.

IF YOU ARE TEACHING a course, a class or a workshop, draw inspiration from existing activities and lesson plans. There are now many excellent eco-emotions guides created by mental health professionals, educators, students, social workers and others that include structured lesson plans and activities you might want to consider using. One of my favorites is "A Brief Guide to Eco-Emotions" (available for free online) created by Ympäristöahdistus, an eco-crisis and mental health project in Finland.[2] You can also find lots of activities on the *Existential Toolkit for Climate Justice Educators* website I helped to co-create with dozens of international researchers and practitioners.

Once you've chosen an activity, consider integrating the following components into the steps in the practice above:

Establish an emotionally safe and trustworthy learning environment

- Like all good relationships, emotional safety is cultivated over time through consistent support and validation of people's thoughts, feelings and needs. Demonstrate compassion and model effective approaches to conflict resolution.

- Build a climate emotions check-in into the beginning and/or end of each class.

- Consider how climate justice issues are currently affecting your life and the lives of the students in the class. Have you or they had (or are they having) direct experiences of fire seasons, floods, heat islands, droughts or severe storms? Are you or they experiencing food or housing insecurity, problems with access to water, health risks associated with air pollution or other indirect experiences of climate change and systemic injustice?

- Let students know that you will be focusing on climate change and that it might be difficult to learn about. Reassure them that the topic of climate change and natural disasters will be approached in a sensitive and respectful manner.

- Use a sensitive content warning at the beginning of the class, or before you share examples of particularly graphic or otherwise disturbing climate justice images or content. You might say: "The content and discussion in this course will necessarily engage with climate change every week. Much of it will be emotionally and intellectually challenging to engage with. I will flag especially graphic or intense content and will do my best to make this classroom a space where we can engage bravely, empathetically and thoughtfully with difficult content every week." Encourage students to pay attention to their feelings. Let them know they can act in their own best interest without ridicule or scrutiny.

- Think about your relationship with respect to power in whatever situation you find yourself. Are you in a position of "power over" the other person (i.e., are you a teacher in conversation with a class of students? An employee in conversation with a volunteer?) Can you identify the impact your power (or powerlessness) might have on the person with whom you are speaking?

- Ask yourself what support you can put in place to strengthen your capacity to hold space for the emotions of others.

- Find out what counseling services are available through your school, university or community. Use them if you need support. Share them with your colleagues and students if they are showing signs of emotional distress.

Structure and teach the entire course from a solutions orientation

- Check your expiry date on the content you are teaching. Consult your solutions journalism sources for up-to-date examples of trends moving in meaningful directions. Pick one or two hopeful examples and weave them into whatever issue you are teaching.

- Create assignments that immerse students in real-world solutions. (Project Drawdown, a leading source for big data–informed climate solutions, has lots of good resources to consider.)

- Repurpose the practice on page 73, "Source solutions-focused news that empowers not harms," as a take-home assignment.

- Use hopeful, age-appropriate language and examples that situate whatever climate issue you're teaching within a wider context of efforts by people in your community or other parts of the world to make positive change. Use specific examples, such as: "Because the city of Seattle made a serious commitment to public transit, it reduced both traffic congestion and greenhouse gas emissions while growing its population by almost a quarter between 2010 and 2020."[3] Avoid catastrophic statements such as "We will never recover from this" or "We are already too far gone."

- Look for ways to teach outside. Are there options to combine the activity with a walk in green space near the school? Strengthening connections in nature builds resilience. Exploring emotions while engaging in light or stress-relieving activities such as gardening or ecological restoration can be a nourishing combination.

HERE'S AN ACTIVITY to invite students to identify their climate emotions.

Pull up the Climate Emotions Wheel, available on the Climate Mental Health Network website (see page 273). Write a variety of the emotions in sidewalk chalk in a parking lot. Invite your students to wander through the words. Ask them to collect as many different words as they need to describe how they are feeling. Create time for individual journal contemplation or sharing in pairs. Encourage reflection. *Was this activity difficult or easy? What, if any, insights did you gain by naming your eco-emotions? Which environmental feelings are easier for you to deal with? Which ones are more difficult?* Invite students to share as much as they might like about their own experience.

Acknowledgments

WRITERS OFTEN CALL OUT a book that could not have been written without the help of others. This sentiment makes me smile, because it is always the case. None of us can write, let alone exist, without the collective generosity of plants who make the oxygen in the air we breathe, soil microbes who nourish the foods we eat, and countless other gifts. A lot of lives make up the multispecies assemblages that we call "me" and "you." I am deeply grateful for these life-giving relationships.

I am indebted to my daughter, Esmé Johnson, for her wisdom and support. Thank you, Esmé, for circling back so many times to remind me of what I was struggling to unlearn and helping me to conceptualize what I was trying to say. The overall structure and feeling of the book finally emerged because of you.

One of the most difficult parts of hope is holding onto the possibility of transformation in the midst of heartache. Thank you, Kip, for loving me so completely that you were willing to go on a reparative quest together. Knowing how wondrous that feels, I am now steadfast in my belief that change can happen no matter what the circumstances.

I am a conceptual thinker who finds details both crucial to making evidence-based arguments and messy to organize on my laptop and in my notebooks. I am full of appreciation for the meticulous and thoughtful care my editor, Paula Ayer, gave to the manuscript. Thank you to Jennifer Stewart for proofreading and for recognizing the value of hope in many situations. Thank you to Fiona Siu, Javana Boothe, and Jess Sullivan for bringing the practices to life through the beautiful design of the book. You would not be reading these words if not for Jen Gauthier (publisher), Megan Jones (sales and marketing director) and all the other thoughtful and enterprising people at Greystone Books.

The proposal for this book was written in a tiny cabin in the California redwoods thanks to a woman named Suzy Stevens who kindly opened her door to me, a complete stranger. That spirit of generosity lived on through all the dear friends, family and other cherished people who cared for me and contributed to the making of this book in so many meaningful ways. I am more thankful to each of you than I can express.

Notes

Unless otherwise noted, all online sources were consulted February 19–21, 2025.

Preface

1   Mukherjee, S., Ray-Mukherjee, J., & Sarabia, R. (2013). Behaviour of American crows (*Corvus brachyrhynchos*) when encountering an oncoming vehicle. *Canadian Field-Naturalist, 127*(3), 229–33. doi.org/10.22621/cfn.v127i3.1488

2   Estimates of the number of species on Earth range from a few million to billions; 8.7 million is one of the most widely cited figures. Mora, C., Tittensor, D. P., Adl, S., Simpson, A. G. B., & Worm, B. (2011). How many species are there on Earth and in the ocean? *PLoS Biology, 9*, e1001127. doi.org/10.1371/journal.pbio.1001127

Introduction

1   Kukla, R. (2018). Embodied stances: Realism without literalism. In B. Huebner (Ed.), *The Philosophy of Daniel Dennett* (pp. 2–35). Oxford Academic. doi.org/10.1093/oso/9780199367511.003.0001

2   Human Rights Council. (2020). *Realizing the rights of the child through a healthy environment.* United Nations doc. A/HRC/43/30, para. 15.

3   United Nations Development Programme. (2024, June 27). *The world's largest survey on climate change is out—here's what the results show.* Climate Promise. climatepromise.undp.org/news-and-stories/worlds-largest-survey-climate-change-out-heres-what-results-show

4   Kenyon, M. (2024, May 10). Michael Mann: "Defeatism is as much of a threat as climate denial." *The New Statesman.* newstatesman.com/spotlight/sustainability/climate/2024/05/michael-mann-defeatism-threat-climate-change-action-net-zero

Stance 1: I Choose Hope

1   Shani, M., Kunst, J. R., Anjum, G., Obaidi, M., ... & Halperin, E. (2024). Between victory and peace: Unravelling the paradox of hope in intractable conflicts. *British Journal of Social Psychology, 63*(3), 1357–84. doi.org/10.1111/bjso.12722

2   Gorvett, Z. (2022, February 24). *How modern life is transforming the human skeleton*. BBC. bbc.com/future/article/20190610-how-modern-life-is-transforming-the-human-skeleton

3   Barcan, R. (2002). Problems without solutions: Teaching theory and the politics of hope. *Continuum, 16*(3), 343–56. doi.org/10.1080/1030431022000018708

4   Style, H. K. (2024). Impression management and expectations of political cynicism. *Public Opinion Quarterly, 88*(2), 419–30. doi.org/10.1093/poq/nfae006

5   Evans, T. (2021, November 3). The myth of the cynical genius. *Psychology Today*. psychologytoday.com/ca/blog/trust-games/202111/the-myth-the-cynical-genius

6   Stavrova, O., & Ehlebracht, D. (2018). The cynical genius illusion: Exploring and debunking lay beliefs about cynicism and competence. *Personality and Social Psychology Bulletin, 45*(2), 254–69. doi.org/10.1177/0146167218783195

7   Zaki, J. (2024, October 7). Instead of being cynical, try becoming skeptical. *Behavioral Scientist*. behavioralscientist.org/instead-of-being-cynical-try-becoming-skeptical

8   Appadurai, A. (2013). Housing and hope. *Places Journal*. doi.org/10.22269/130305

9   Stockdale, K. (2011). Hope and anger. In *Hope under oppression* (pp. 82–113). Oxford University Press. doi.org/10.1093/oso/9780197563564.003.0004

10   TEDx Talks. (2015, May 21). Anger is not a bad word | Myisha Cherry | TEDxUofIChicago [Video]. YouTube. youtube.com/watch?v=uysTk2E10tw

11   Troy, C. (2024, December 3). *How solutions journalism can transform climate change reporting and inspire action*. PennState Institute of Energy and the Environment. iee.psu.edu/news/blog/how-solutions-journalism-can-transform-climate-change-reporting-and-inspire-action

12   Gnanasambandan, S. (2023, October 6). The secret to a long life [Audio podcast episode]. In *Radiolab*. WNYC Studios. radiolab.org/podcast/secret-long-life

13   Fennell, L. A. (2024, June 12). What shape does progress take? Don't assume it's a straight line. *Behavioral Scientist*. behavioralscientist.org/what-shape-does-progress-take-dont-assume-its-a-straight-line

14   The Annie E. Casey Foundation. (2024, February 15). *A new way to measure healing from violence* [Webinar]. The Annie E. Casey Foundation Blog. aecf.org/blog/a-new-way-to-measure-healing-from-violence

15   King, M. L., Jr. (1962–1963). *Draft of chapter X, "Shattered dreams."* The Martin Luther King, Jr. Research and Education Institute, Stanford University. kinginstitute.stanford.edu/king-papers/documents/draft-chapter-x-shattered-dreams

16   Kopecky, A. (2024, July 15). The world is moving away from fossil fuels. Canada is holding on for dear life. *The Walrus*. thewalrus.ca/the-world-is-moving-away-from-fossil-fuels-canada-is-holding-on-for-dear-life

17   Bloch, E. (1995). *The principle of hope* (3 vols). MIT Press; Huber, J. (2019). Defying democratic despair: A Kantian account of hope in politics. *European Journal of Political Theory, 20*(4), 719–38. doi.org/10.1177/1474885119847308

18   Aschwanden, C. (2024, April 3). Uncertainty is science's superpower. Make it yours, too [Audio podcast episode]. In *Uncertain*. Scientific American. scientificamerican.com/podcast/episode/uncertainty-is-sciences-super-power-make-it-yours-too

19 Hester, A. (2017, June 29). *Mapping hope in the Israel-Palestine conflict*. Peace Insight. peaceinsight.org/en/articles/mapping-hope-israel-palestine-conflict

20 Leshem, O. A., & Halperin, E. (2020). Hope during conflict. In S. C. van den Heuvel (Ed.), *Historical and multidisciplinary perspectives on hope* (pp. 179–96). SpringerOpen. doi.org/10.1007/978-3-030-46489-9_10

21 Moeschberger, S. L., Dixon, D. N., Niens, U., & Cairns, E. (2005). Forgiveness in Northern Ireland: A model for peace in the midst of the "Troubles." *Peace and Conflict: Journal of Peace Psychology, 11*(2), 199–214. doi.org/10.1207/s15327949paci1102_5

22 Laverne, L. (2022, February 4). Lyse Doucet, journalist [Audio podcast episode]. In *Desert Island Discs*. BBC Radio 4. bbc.co.uk/programmes/m0013yzs

23 United Nations Development Programme. (2022, February 8). *2022 special report on human security*. UNDP. hdr.undp.org/content/2022-special-report-human-security

24 Taylor, A., & Ramamurthy, R. (2023, September 19). Capitalism, the insecurity machine: A conversation with Astra Taylor. *Nonprofit Quarterly*. nonprofitquarterly. org/capitalism-the-insecurity-machine-a-conversation-with-astra-taylor

25 Bigoni, M., & Mohammed, S. (2023). Critique is unsustainable: A polemic. *Critical Perspectives on Accounting 97*, 102555. doi.org/10.1016/j.cpa.2023.102555

Stance 2: I Reject Fatalism

1 Sauer, B. (2023). The in_visibilization of emotions in politics. Ambivalences of an "affective democracy." *Journal of Gender Studies, 32*(8), 819–31. doi.org/10.1080/09589236.2023.2227116

2 Simpkins, K. (2021, December 21). Climate change news coverage reached all-time high, language to describe it shifting. *CU Boulder Today*. colorado.edu/today/2021/12/21/climate-change-news-coverage-reached-all-time-high-language-describe-it-shifting

3 Saab, A. (2023). Discourses of fear on climate change in international human rights law. *European Journal of International Law, 34*(1), 113–35. doi.org/10.1093/ejil/chad002

4 Hulme, M. (2006, November 4). *Chaotic world of climate truth*. BBC News. news.bbc.co.uk/2/hi/science/nature/6115644.stm

5 Altheide, D. L. (2016). Media culture and the politics of fear. *Culture Studies ↔ Critical Methodologies, 19*(1), 3–4. doi.org/10.1177/1532708616655749

6 Chaiuk, T., & Dunaievska, O. V. (2020, June 24). Fear culture in media: An examination on coronavirus discourse. *Journal of History Culture and Art Research, 9*(2), 184. researchgate.net/publication/342823673

7 WVU study finds control, fear and shame tactics don't work for effective messaging. (2022, September 8). *WVU Today*. wvutoday.wvu.edu/stories/2022/09/08/wvu-study-finds-control-fear-and-shame-tactics-don-t-work-for-effective-messaging

8 Gardner, K. (2023, April 5). *Climate change has a branding problem, here's how we fix that*. Constructive. constructive.co/insight/positive-climate-communications

9 Swim, J., Clayton, S., Doherty, T., Gifford, R., ... & Weber, E. (2010). *Psychology and global climate change: Addressing a multi-faceted phenomenon and set of challenges*. A report by the American Psychological Association's Task Force on the Interface

Between Psychology and Global Climate Change. apa.org/science/about/publications/climate-change.pdf

10 Vigna, L., Friedrich, J., & Damassa, T. (2024, June 3). *The history of carbon dioxide emissions.* World Resources Institute. wri.org/insights/history-carbon-dioxide-emissions

11 Environment and Climate Change Canada. (2024). *Canadian environmental sustainability indicators: Greenhouse gas emissions.* canada.ca/content/dam/eccc/documents/pdf/cesindicators/ghg-emissions/2024/greenhouse-gas-emissions-en.pdf

12 Niranjan, A. (2024, October 31). EU emissions fall by 8% in steep reduction reminiscent of Covid shutdown. *The Guardian.* theguardian.com/environment/2024/oct/31/eu-emissions-fall-by-8-in-steep-reduction-reminiscent-of-covid-shutdown

13 Mayer, A., & Keith Smith, E. (2019). Unstoppable climate change? The influence of fatalistic beliefs about climate change on behavioural change and willingness to pay cross-nationally. *Climate Policy, 19*(4), 511–23. doi.org/10.1080/14693062.2018.1532872

14 Borenstein, S. (2022, April 5). Earth needs climate action, not climate "doomism," scientists say. *The Christian Science Monitor.* csmonitor.com/Environment/2022/0405/Earth-needs-climate-action-not-climate-doomism-scientists-say

15 DiGirolamo, M., & Donald, R. (2025, January 7). Former UN climate chief Christiana Figueres remains optimistic despite disappointing COP process [Audio podcast episode]. In *Mongabay Newscast.* Mongabay. news.mongabay.com/podcast/former-un-climate-chief-christiana-figueres-remains-optimistic-despite-disappointing-cop-process

16 Filkowski, M. M., Cochran R. N., & Haas, B. W. (2016). Altruistic behavior: Mapping responses in the brain. *Neuroscience and Neuroeconomics, 5,* 65–75. doi.org/10.2147/NAN.S87718

17 Manny's. (2022, March 24). The practice and power of forgiving w/ Dr Fred Luskin [Video]. YouTube. youtube.com/watch?v=tzAdY-608Ds

18 Katz, R. (2023, October 23). The enemies of gratitude [Audio podcast episode]. In *Hidden Brain.* Hidden Brain Media. hiddenbrain.org/podcast/the-enemies-of-gratitude

19 Nuccitelli, D. (2023, November 27). *Most people don't realize how much progress we've made on climate change.* Yale Climate Connections. yaleclimateconnections.org/2023/11/most-people-dont-realize-how-much-progress-weve-made-on-climate-change

20 Harvey, F. (2023, March 20). Scientists deliver "final warning" on climate crisis: Act now or it's too late. *The Guardian.* theguardian.com/environment/2023/mar/20/ipcc-climate-crisis-report-delivers-final-warning-on-15c

21 Newman, N., Fletcher, R., Schultz, A., Andi, S., & Nielsen, R. K. (2020). *Digital news report 2020.* Reuters Institute for the Study of Journalism. doi.org/10.60625/risj-048n-ap07; Lynas, M., Houlton, B. Z., & Perry, S. (2021). Greater than 99% consensus on human caused climate change in the peer-reviewed scientific literature. *Environmental Research Letters, 16*(11), 114005. doi.org/10.1088/1748-9326/ac2966; Bretter, C., & Schulz, F. (2023). Why focusing on "climate change denial" is counterproductive. *Proceedings of the National Academy of Sciences, 120*(10), e2217716120. doi.org/10.1073/pnas.2217716120

22 Andre, P., Boneva, T., Chopra, F., & Falk, A. (2024). Globally representative evidence on the actual and perceived support for climate action. *Nature Climate Change, 14*(3), 253–59. doi.org/10.1038/s41558-024-01925-3

23 Sparkman, G., Geiger, N., & Weber, E. U. (2022). Americans experience a false social reality by underestimating popular climate policy support by nearly half. *Nature Communications, 13*(1), 4779. doi.org/10.1038/s41467-022-32412-y

24 Rose, T. (2022). *Collective illusions: Conformity, complicity, and the science of why we make bad decisions.* Hachette, 6.

25 *New research concludes that climate framework laws have had a positive impact on climate action in New Zealand, Germany and Ireland.* (2024, March 14). Grantham Research Institute on Climate Change and the Environment. lse.ac.uk/granthaminstitute/news/new-research-concludes-that-climate-framework-laws-have-had-a-positive-impact-on-climate-action-in-new-zealand-germany-and-ireland

26 Hase, V., Mahl, D., Schäfer, M. S., & Keller, T. R. (2021). Climate change in news media across the globe: An automated analysis of issue attention and themes in climate change coverage in 10 countries (2006–2018). *Global Environmental Change, 70*, 102353. doi.org/10.1016/j.gloenvcha.2021.102353

27 Khamaiseh, M. (2023, September 11). *How do we determine "newsworthiness" in the digital age?* Al Jazeera Media Institute. institute.aljazeera.net/en/ajr/article/2329

28 Perga, M.-E., Pessina, L.-A., Lane, S., & Butera, F. (2024, February 7). From newsworthiness to news usefulness in climate change research. *Eos, 105.* doi.org/10.1029/2024EO240051; Perga, M.-E., Sarrasin, O., Steinberger, J., Lane, S. N. & Butera, F. (2023). The climate change research that makes the front page: Is it fit to engage societal action? *Global Environmental Change, 80*, 102675, doi.org/10.1016/j.gloenvcha.2023.102675

29 Saab, A. (2003). Discourses of fear on climate change in international human rights law. *European Journal of International Law, 34*(1), 113–35, doi.org/10.1093/ejil/chad002

30 Perga, M.-E., Pessina, L.-A., Lane, S., & Butera, F. (2024, February 7). From newsworthiness to news usefulness in climate change research. *Eos, 105.*

31 van der Meer, T. G. L. A., & Hameleers, M. (2022). I knew it, the world is falling apart! Combatting a confirmatory negativity bias in audiences' news selection through news media literacy interventions. *Digital Journalism, 10*(3), 473–92. doi.org/10.1080/21670811.2021.2019074

32 Kassova, L. (2022). *Is journalism inadvertently contributing to climate inaction?* World Association of News Publishers. wan-ifra.org/2022/02/is-journalism-inadvertently-contributing-to-climate-inaction; *Don't feed fatalism . . . put forward solutions instead.* (2020, June 6). FrameWorks Institute. frameworksinstitute.org/article/dont-feed-fatalism-put-forward-solutions-instead

33 Ricchiardi, S. (2023, December 4). *Climate change through a solutions and data lens.* DataJournalism.com. datajournalism.com/read/longreads/climate-change-through-a-solutions-and-data-lens

34 Achor, S., & Gielan, M. (2015, September 14). Consuming negative news can make you less effective at work. *Harvard Business Review.* hbr.org/2015/09/consuming-negative-news-can-make-you-less-effective-at-work

35 Gorvett, Z. (2020, May 12). *How the news changes the way we think and behave.* BBC. bbc.com/future/article/20200512-how-the-news-changes-the-way-we-think-and-behave

36 Holman, E. A., Garfin, D. R., & Silver, R. C. (2024). It matters what you see: Graphic media images of war and terror may amplify distress. *Proceedings of the National Academy of Sciences, 121*(29), e2318465121. doi.org/10.1073/pnas.2318465121

37 Blades, R. (2021). Protecting the brain against bad news. *Canadian Medical Association Journal, 193*(12), E428–29. doi.org/10.1503/cmaj.1095928

38 Tomhave, K. (2020, June 12). *Re-thinking the poverty narrative: Journalists, academics should highlight who benefits from inequality.* Poverty Solutions at the University of Michigan. poverty.umich.edu/2020/06/12/re-thinking-the-poverty-narrative-journalists-academics-should-highlight-who-benefits-from-inequality

39 Averbuch, M. (2024, October 14). How Mexico City averted all-out drought. *Bloomberg.* bloomberg.com/news/features/2024-10-14/mexico-city-day-zero-never-came-how-the-city-avoided-running-out-of-water

40 Newman, N., Fletcher, R., Schultz, A., Andi, S., & Nielsen, R. K. (2020). *Digital news report 2020.* Reuters Institute for the Study of Journalism. doi.org/10.60625/risj-048n-ap07; Betakova, D., Boomgaarden, H., Lecheler, S., & Schäfer, S. (2024). I do not (want to) know! The relationship between intentional news avoidance and low news consumption. *Mass Communication and Society,* 1–28. doi.org/10.1080/15205436.2024.2304759

41 Longpré, C., Sauvageau, C., Cernik, R., Journault, A.-A., ... & Lupien, S. Staying informed without a cost: No effect of positive news media on stress reactivity, memory and affect in young adults. *PLoS One, 16*(10), e0259094. doi.org/10.1371/journal.pone.0259094

42 Linnitt, C. (2025, January 23). There's no point in despair when there's so much to hope for. *The Narwhal.* thenarwhal.ca/newsletter-remain-hopeful-political-changes

43 TEDx Talks. (2024, December 1). How to save our children from cynicism | Dr. Graeme Mitchell | TEDxRRU [Video]. YouTube. youtube.com/watch?v=sPcyUd9lbyc

44 Upton, J. (2015, March 28). *Media contributing to "hope gap" on climate change.* Climate Central. climatecentral.org/news/media-hope-gap-on-climate-change-18822

45 Doll, S. (2024). *A fully-electric 10,000 ton container ship has begun service equipped with over 50,000 kWh in batteries.* Electrek. electrek.co/2024/05/02/fully-electric-10000-ton-container-ship-begun-service50000-kwh-batteries; Lakin, T. (2023). Climate-quitting: What is it and why it matters to HR leaders. *HR.* hrmagazine.co.uk/content/comment/climate-quitting-what-is-it-and-why-it-matters-to-hr-leaders; Washington State Department of Transportation. (n.d.). *Federal court injunction for fish passage.* wsdot.wa.gov/construction-planning/protecting-environment/fish-passage/federal-court-injunction-fish-passage

46 Frost, R. (2024, May 8). More than 30% of world's electricity now comes from renewables, report reveals. *Euronews.* euronews.com/green/2024/05/08/a-major-turning-point-more-than-30-of-worlds-energy-now-comes-from-renewables-report-revea; Carrington, D. (2024, January 5). African elephant populations stabilise in southern heartlands. *The Guardian.* theguardian.com/environment/2024/jan/05/african-elephant-populations-stabilise-in-southern-heartlands

47 Gapminder, gapminder.org

Stance 3: I Am Emotional

1 Lerner, J. S., Li, Y., Valdesolo, P., & Kassam, K. S. (2015). Emotion and decision making. *Annual Review of Psychology, 66*, 799–823. doi.org/10.1146/annurev-psych-010213-115043

2 Moser, S. C. (2020). The adaptive mind. In A. E. Johnson & K. K. Wilkinson (Eds.), *All we can save: Truth, courage, and solutions for the climate crisis* (pp. 270–78). One World.

3 Pihkala, P. (2020). Anxiety and the ecological crisis: An analysis of eco-anxiety and climate anxiety. *Sustainability, 12*(19), 7836. doi.org/10.3390/su12197836

4 Anne Ylvisaker, anneylvisaker.com

5 Diniz, C. R. A. F., & Crestani, A. P. (2023). The times they are a-changin': A proposal on how brain flexibility goes beyond the obvious to include the concepts of "upward" and "downward" to neuroplasticity. *Molecular Psychiatry, 28*(3), 977–92. doi.org/10.1038/s41380-022-01931-x

6 Feldman Barrett, L. (2024). The "fight or flight" idea misses the beauty of what the brain really does. *Scientific American.* scientificamerican.com/article/simplistic-fight-or-flight-idea-undervalues-the-brains-predictive-powers; Callahan, M. (2019). *It's time to correct neuroscience myths.* Northeastern University College of Science. cos.northeastern.edu/news/its-time-to-correct-neuroscience-myths

7 Carey, L. (2024, April 8). Reptiles are highly emotional, contrary to their cold reputation. *Discover Magazine.* discovermagazine.com/planet-earth/reptiles-are-highly-emotional-contrary-to-their-cold-reputation

8 Alexander, R., Aragón, O. R., Bookwala, J., Cherbuin, N., ... & Styliadis, C. (2021). The neuroscience of positive emotions and affect: Implications for cultivating happiness and wellbeing. *Neuroscience & Biobehavioral Reviews, 121*, 220–49, doi.org/10.1016/j.neubiorev.2020.12.002

9 Henrich, J., Heine, S. J., & Norenzayan, A. (2010). The weirdest people in the world? *Behavioral and Brain Sciences, 33*(2–3): 61–83. doi.org/10.1017/S0140525X0999152X

10 Apicella, C., Norenzayan, A., & Henrich, J. (2020). Beyond WEIRD: A review of the last decade and a look ahead to the global laboratory of the future. *Evolution and Human Behavior, 41*(5), 319–29. doi.org/10.1016/j.evolhumbehav.2020.07.015

11 Morgan, E., & Zahl, B. (2021, October 15). *Beyond WEIRD: Why we need to make psychology and social science research more inclusive.* Templeton World Charity Foundation, Inc. templetonworldcharity.org/blog/beyond-weird-why-we-need-make-psychology-and-social-science-research-more-inclusive

12 *Cultural and psychological distance: Beyond weirdness.* (2021, August 30). Association for Psychological Science. psychologicalscience.org/observer/cultural-distance

13 McCaffery, J., & Boetto, H. (2024). Eco-emotional responses to climate change: A scoping review of social work literature. *The British Journal of Social Work, 55*(1), 120–40. doi.org/10.1093/bjsw/bcae129

14 Ray, S. J. (2021). Who feels climate anxiety? *The Cairo Review of Global Affairs.* thecairoreview.com/essays/who-feels-climate-anxiety

15 Qiu, S., & Qiu, J. (2024). From individual resilience to collective response: Reframing ecological emotions as catalysts for holistic environmental engagement. *Frontiers in Psychology, 15*, 1363418. doi.org/10.3389/fpsyg.2024.1363418

16   Montgomery, E. (2024, September 20). *Americans deliver 1 million comments in favor of old-growth forests*. Environment America. environmentamerica.org/media-center/americans-deliver-1-million-comments-in-favor-of-old-growth-forests

17   Waters, S. (2024, December 13). B.C. government aims to permanently protect Fairy Creek. *The Narwhal*. thenarwhal.ca/bc-fairy-creek-protection-pledge

18   Pihkala, P. (2022). The process of eco-anxiety and ecological grief: A narrative review and a new proposal. *Sustainability, 14*(24), 16628. doi.org/10.3390/su142416628

19   Process of Eco-Anxiety. (2024, November 25). The dance: Living with eco-anxiety [Video]. YouTube. youtube.com/watch?v=_Jexuu4DWao

20   Rackete, C. (2021, December 2). What privilege means in the climate crisis fight. *The New York Times*. nytimes.com/2021/12/02/special-series/climate-crisis-responsibility-privilege.html

21   Feldman Barrett, L. (2018, May/June). The science of making emotions. *Healthy Living Made Simple*, 38–39. lisafeldmanbarrett.com/wp-content/uploads/sites/4/2020/11/ScienceOfMakingEmotions.pdf

22   Armstrong, K. (2020, January 29). *Remarkable resiliency: George Bonanno on PTSD, Grief, and Depression*. Association for Psychological Science. psychologicalscience.org/observer/bonanno

23   Kozubal, M., Szuster, A., & Wielgopolan, A. (2023). Emotional regulation strategies in daily life: The intensity of emotions and regulation choice. *Frontiers in Psychology, 14*, 1218694. doi.org/10.3389/fpsyg.2023.1218694

24   Welby, J. (2023, June 18). Zarifa Ghafari [Audio podcast episode]. In *The Archbishop Interviews*. BBC. bbc.co.uk/sounds/play/m001n1vs

25   Feldman Barrett, L. (2018, May/June). The science of making emotions. *Healthy Living Made Simple*, 38–39.

Stance 4: I Am on a Reparative Quest of Transformation

1   National Centre for Truth and Reconciliation. (n.d.). *Residential school history*. NCTR. nctr.ca/education/teaching-resources/residential-school-history

2   National Oceanic and Atmospheric Administration Fisheries. (2024, October 29). *Saving the southern resident killer whales*. NOAA Fisheries. fisheries.noaa.gov/west-coast/endangered-species-conservation/saving-southern-resident-killer-whales

3   Hickman, C., Marks, E., Pihkala, P., Clayton, S., ... & van Susteren, L. (2021). Climate anxiety in children and young people and their beliefs about government responses to climate change: A global survey. *The Lancet Planetary Health 5*(12), e863–73, doi.org/10.1016/s2542-5196(21)00278-3

4   Jones, C. A., & Davison, A. (2021). Disempowering emotions: The role of educational experiences in social responses to climate change. *Geoforum, 118*, 190–200. doi.org/10.1016/j.geoforum.2020.11.006

5   World's Youth for Climate Justice, Pacific Islands Students Fighting Climate Change, Sobenes, E., Alarcon, M. J., & Rose, J. (2023). *The youth climate justice handbook—Part 1: Summary for policymakers*. WYCJ and PISFCC. wy4cj.org/handbook

6   Heilman, E. E. (2022). Anger is all the rage: A theoretical analysis of anger within emotional ecology to foster growth and political change. *Teachers College Record, 124*(4), 205–34. doi.org/10.1177/01614681221093285

7   Kaji, L. B. (2022, November 30). Not seeing red: An exploration of anger in activism. *Glasgow University Magazine*. glasgowuniversitymagazine.co.uk/articles/features/not-seeing-red-an-exploration-of-anger-in-activism

8   Mishra, P. (2017). *Age of anger: A history of the present*. Farrar, Straus and Giroux, 27.

9   Keren, R. (2016, March 23). *Nussbaum: Anger is the wrong response to injustice*. Middlebury News and Announcements. www.middlebury.edu/announcements/news/2016/03/nussbaum-anger-wrong-response-injustice

10   Gandhi, M. K., Gandhi, K., & Surabati, A. (Eds.). (1920, September 15). *Young India: A Weekly Journal, 2*(37).

11   Mandela, N. (2011). *Conversations with myself*. Anchor Canada.

12   Palmater, P. (2019, April 6). Cindy Blackstock on justice and equality for First Nations children [Video]. YouTube. youtube.com/watch?v=t5LVH3LY_20

13   Surian, L., & Margoni, F. (2020). Commentary: Children's sense of fairness as equal respect. *Frontiers in Psychology, 11*. doi.org/10.3389/fpsyg.2020.00107

14   Ober, H. (2023). *Inequality not inevitable among mammals, study shows*. UCLA Newsroom. newsroom.ucla.edu/releases/study-shows-inequality-not-inevitable-among-mammals; Smith, J. E., Natterson-Horowitz, B., Mueller, M. M., & Alfaro, M. E. (2023). Mechanisms of equality and inequality in mammalian societies. *Philosophical Transactions of the Royal Society B, 378*(1883), 20220307. doi.org/10.1098/rstb.2022.0307

15   Deutsches Primatenzentrum (DPZ)/German Primate Center. (2023, March 23). Insights into the evolution of the sense of fairness. *ScienceDaily*. sciencedaily.com/releases/2023/03/230302114205.htm

16   Blakeley, G. (2024, August 28). Capitalism's gaping inequalities are also its main weakness—and the spur for resistance. *LSE Inequalities*. blogs.lse.ac.uk/inequalities/2024/08/28/capitalisms-gaping-inequalities-are-also-its-main-weakness

17   Romero, G. (2024, October 2). *Recommendations for universities worldwide for the second semester of 2024: Safeguarding the right to freedom of peaceful assembly and association on campuses in the context of international solidarity with the Palestinian people and victims*. United Nations Human Rights Special Procedures: Statement from the UN Special Rapporteur on the Rights of Freedom of Peaceful Assembly and of Association. ohchr.org/sites/default/files/documents/issues/association/statements/20241004-stm-sr-association.pdf

18   Noor, D. (2024, April 24). How divestment became a "clarion call" in anti–fossil fuel and pro-ceasefire protests. *The Guardian*. theguardian.com/us-news/2024/apr/24/university-fossil-fuel-divestment-student-protests-israel-gaza

19   Ober, H. (2021, November 2). Anger as an appropriate power source for social justice. *UC Riverside News*. news.ucr.edu/articles/2021/11/02/anger-appropriate-power-source-social-justice

20   Schwartz, J. A., Lendway, P., & Nuri, A. (2023). Fossil fuel divestment and public climate change policy preferences: An experimental test in three countries. *Environmental Politics, 33*(1), 1–24. doi.org/10.1080/09644016.2023.2178351

21   Freyd, J. J. (2024, March 10). *Institutional betrayal and institutional courage*. Jennifer J. Freyd, PhD. dynamic.uoregon.edu/jjf/institutionalbetrayal

22   Freyd, J. J. (n.d.). *What is DARVO?* Jennifer Joy Freyd, PhD. jjfreyd.com/darvo

23   Romero, G. (2024, October 2). *Recommendations for universities worldwide for the second semester of 2024: Safeguarding the right to freedom of peaceful assembly and association on campuses in the context of international solidarity with the Palestinian people and victims.* United Nations Human Rights Special Procedures: Statement from the UN Special Rapporteur on the Rights of Freedom of Peaceful Assembly and of Association, 9.

24   Center for Institutional Courage. (n.d.). *We are making a call to institutional courage.* Center for Institutional Courage. institutionalcourage.org/the-call-to-courage

25   Million, D. (2009). Felt theory: An Indigenous feminist approach to affect and history. *Wicazo Sa Review 24*(2), 53–76. doi.org/10.1353/wic.0.0043

26   Joshi, Y. (2023, June). Weaponizing peace. *Columbia Law Review, 123*(5), 1411–48. columbialawreview.org/content/weaponizing-peace

27   Blakeley, G. (2024, October 22). Capitalism's gaping inequalities are also its main weakness—and the spur for resistance. *LSE Inequalities.*

28   Slaby, J. (2023). Structural apathy, affective injustice, and the ecological crisis. *Philosophical Topics, 51*(1), 63–84. doi.org/10.5840/philtopics20235114

29   Almassi, B. (2021). *Philosophers as advocates for reparative climate justice.* Philosophers for Sustainability Conference. philosophersforsustainability.com/wp-content/uploads/2021/04/2a-almassi-philosophers-as-advocates-for-reparative-climate-justice.pdf

30   Wilkins, E. (2024, December 18). *Beyond forgiveness: The reparative quest in South Africa.* John Templeton Foundation. templeton.org/news/beyond-forgiveness-the-reparative-quest-in-south-africa

31   Duhamel, K. (n.d.). *Reconciliation: A movement of hope or a movement of guilt?* Canadian Museum for Human Rights. humanrights.ca/story/reconciliation-movement-hope-or-movement-guilt

32   Joseph Rowntree Foundation. (2023, September 15). Hospicing modernity: Vanessa Andreotti [Video]. YouTube. youtube.com/watch?v=x0zk0FFmbIY

33   Popova, M. (2023, October 22). 17 life-learnings from 17 years of *The Marginalian.* *The Marginalian.* themarginalian.org/2023/10/22/17

34   Sachs, J. D. (1999, December 1). Twentieth-century political economy: A brief history of global capitalism. *Oxford Review of Economic Policy, 15*(4), 90–101. doi.org/10.1093/oxrep/15.4.90

35   Green European Journal Staff & Hickel, J. (2022, January 7). *Degrowth is about global justice.* Resilience. resilience.org/stories/2022-01-07/degrowth-is-about-global-justice

36   Kallis, G., Kostakis, V., Lange, S., Muraca, B., … & Schmelzer, M. (2018). Research on degrowth. *Annual Review of Environment and Resources, 43*(1), 291–316. doi.org/10.1146/annurev-environ-102017-025941

37   Gilbert, D. (2014). The psychology of your future self [Video]. TED Talks. ted.com/talks/dan_gilbert_the_psychology_of_your_future_self

38   Centre for Teaching and Learning. (n.d.). *Positionality statement.* Queen's University. queensu.ca/ctl/resources/equity-diversity-inclusivity/positionality-statement

39   Graham, S. (2016). A review of *Settler Identity and Colonialism in 21st Century Canada*, by Emma Battell Lowman and Adam J. Barker. *[Power and Identity] in Education, 22*(2), 98–100. doi.org/10.37119/ojs2016.v22i2.327

40   McGuire-Adams, T. (2021). Settler allies are made, not self-proclaimed: Unsettling conversations for non-Indigenous researchers and educators involved in Indigenous health. *Health Education Journal, 80*(7), 761–72. doi.org/10.1177/0017896921109269

41   Duhamel, K. (n.d.). *Reconciliation: A movement of hope or a movement of guilt?* Canadian Museum for Human Rights.

42   Jackson, S., & Humphrey, C. (2022, July 28). *Yale experts explain intersectionality and climate change.* Yale Sustainability. sustainability.yale.edu/explainers/yale-experts-explain-intersectionality-and-climate-change

43   Capshaw-Mack, S. (2021, November 10). *A conversation with Leah Thomas, intersectional environmentalist.* State of the Planet—News from the Columbia Climate School. news.climate.columbia.edu/2021/11/10/a-conversation-with-leah-thomas-intersectional-environmentalist

44   Mehri, M. (2020, July 7). Anti-racism requires so much more than "checking your privilege." *The Guardian.* theguardian.com/commentisfree/2020/jul/07/anti-racism-checking-privilege-anti-blackness; Creative Equity Toolkit. (n.d.). *Educate yourself.* Diversity Arts Australia and The British Council. creativeequitytoolkit.org/topic/anti-racism/educate-yourself

45   Frischmann, C., & Chissell, C. (2021, October 27). *The powerful role of household actions in solving climate change.* Project Drawdown. drawdown.org/insights/the-powerful-role-of-household-actions-in-solving-climate-change

46   Yale Forum on Religion and Ecology. (n.d.). *Climate change statements from world religions.* Yale School of the Environment. fore.yale.edu/climate-emergency/climate-change-statements-from-world-religions

47   Faith Pavilion. (2023, November 23). *Global faith leaders unite to declare support.* faithatcop28.com/global-faith-leaders-unite-to-declare-support

48   Roman Catholic Archdiocese of Washington. (n.d.). *The year of the Jubilee 2025: Pilgrims of hope.* adw.org/the-year-of-the-jubilee-2025-pilgrims-of-hope

49   Gayle, D. (2023, July 31). Supermarket plastic bag charge has led to 98% drop in use in England, data shows. *The Guardian.* theguardian.com/environment/2023/jul/31/government-urged-to-repeat-success-of-plastic-bag-charge

50   Bega, S. (2024, April 18). South Africans agree with global consensus that single-use plastics should be banned. *The Mail & Guardian.* mg.co.za/news/2024-04-17-south-africans-agree-with-global-consensus-that-single-use-plastics-should-be-banned

51   Fabra, N., Pintassilgo, C., & Souza, M. (2024). Observed patterns of free-floating car-sharing use. *SERIEs, 15*(3), 259–97. doi.org/10.1007/s13209-024-00298-2

52   Fenes, G. (2023, December 27). *Carpooling: These are the results of the plan promoting sustainable mobility in France.* Mobility Portal Europe. mobilityportal.eu/carpooling-plan-sustainable-mobility-france

53   Mayor of London, London Assembly. (2024, March 9). *New report reveals dramatic improvements in London's air quality since 2016.* Greater London Authority. london.gov.uk/new-report-reveals-dramatic-improvements-londons-air-quality-2016

54   Anderson, R., & Zanger, J. (2025, January 5). *Congestion pricing starts Jan. 5 in New York City. Here's what to know about the new timeline*. CBS News. cbsnews.com/newyork/news/nyc-congestion-pricing-new-start-date

55   Staley, O. (2024, June 7). As New York retreats from charging drivers more, these cities are pushing ahead. *Washington Post*. washingtonpost.com/climate-solutions/2024/06/06/new-york-city-congestion-pricing-london-stockholm

56   Laverne, L. (2022, June 26). Bono, singer and songwriter [Audio podcast episode]. In *Desert Island Discs*. BBC Radio 4. bbc.co.uk/programmes/m0018njm

57   Popova, M. (2024, October 22). 18 life-learnings from 18 years of *The Marginalian*. *The Marginalian*. themarginalian.org/2024/10/22/marginalian-18

58   Carlson, B., & Burnett, T. (2024, June 4). *Dr. Pumla Gobodo-Madikizela receives 2024 Templeton Prize*. John Templeton Foundation. templeton.org/news/dr-pumla-gobodo-madikizela-receives-2024-templeton-prize

59   Tutu, D. M. (2003). No future without forgiveness [Speech]. Archbishop Desmond Tutu Collection, University of North Florida. digitalcommons.unf.edu/archbishoptutupapers/6

60   Tutu, D. M. (2000). *No future without forgiveness*. Image, 205.

Stance 5: I Am Nature

I am grateful to the Rachel Carson Center for Environment and Society for so many things, including granting me permission to use the text describing why I sleep outside, which was originally published in *Springs*. Kelsey, E. (2022, December 13). Why I sleep outside. *Springs: The Rachel Carson Center Review, 2*. doi.org/10.5282/RCC-SPRINGS-2881

1   Tourula, M., Isola, A., & Hassi, J. (2008). Children sleeping outdoors in winter: Parents' experiences of a culturally bound childcare practice. *International Journal of Circumpolar Health, 67*(2–3), 269–78. doi.org/10.3402/ijch.v67i2-3.18284

2   Affifi, R. (2023). Aesthetic knowing and ecology: Cultivating perception and participation during the ecological crisis. *Environmental Education Research, 30*(7), 1041–60. doi.org/10.1080/13504622.2023.2286933

3   Wildcat, D. R. (2013). Introduction: Climate change and Indigenous Peoples of the USA. *Climatic Change, 120*, 509–15. doi.org/10.1007/s10584-013-0849-6. Cited in Dieckmann, U. (2023). Thinking with relations in nature conservation? A case study of the Etosha National Park and Haiǁom. *Journal of the Royal Anthropological Institute, 29*(4), 859–79. doi.org/10.1111/1467-9655.14008

4   McGregor, D. (2015). Indigenous women, water justice and *Zaagidowin* (love). *Canadian Woman Studies/Les cahiers de la femme, 30*(2–3), 71–78. cws.journals.yorku.ca/index.php/cws/article/view/37455/34003

5   Lende, D. (2010, July 7). *Squirrels as models for human behavior? Indeed!* Neuroanthropology. neuroanthropology.net/2010/07/07/squirrels-as-models-for-human-behavior-indeed

6   Garvey, K. K. (2022, June 21). *Rick Karban research: Do plants have personalities?* Entomology & Nematology News, UC ANR. ucanr.edu/blogs/blogcore/postdetail.cfm?postnum=53490

7   Baluška, F., & Mancuso, S. (2020). Plants, climate and humans: Plant intelligence changes everything. *EMBO reports, 21*(3), e50109. doi.org/10.15252/embr.202050109

8   Leonard, P. (2015, October 7). *A new flyway: Fall migrants cross the Atlantic to reach South America*. All About Birds, Cornell Lab of Ornithology. allaboutbirds.org/ news/a-new-flyway-fall-migrants-cross-the-atlantic-to-reach-south-america

9   Drymon, J. M., Feldheim, K., Fournier, A. M. V., Seubert, E. A., ... & Powers, S. P. (2019). Tiger sharks eat songbirds: Scavenging a windfall of nutrients from the sky. *Ecology, 100*(9), e02728. doi.org/10.1002/ecy.2728

10   Varanasi, A. (2022, September 21). *How colonialism spawned and continues to exacerbate the climate crisis*. State of the Planet—News from the Columbia Climate School. news.climate.columbia.edu/2022/09/21/how-colonialism-spawned-and-continues-to-exacerbate-the-climate-crisis

11   Varanasi, A. (2022, September 21). *How colonialism spawned and continues to exacerbate the climate crisis*. State of the Planet—News from the Columbia Climate School.

12   Ellis-Petersen, H. (2022, January 14). Amitav Ghosh: European colonialism helped create a planet in crisis. *The Guardian*. theguardian.com/books/2022/jan/14/ amitav-ghosh-european-colonialism-helped-create-a-planet-in-crisis

13   Smith, R. (2018, January 15). *First Nations leading land and marine stewardship in the Great Bear Rainforest and Sea*. Environmental Fellows Program, Yale School of the Environment. environmentalfellows.yale.edu/reflections/first-nations-leading-land-and-marine-stewardship-great-bear-rainforest-and-sea

14   Roy, R. D. (2018, April 5). Decolonise science—time to end another imperial era. *The Conversation*. theconversation.com/decolonise-science-time-to-end-another-imperial-era-89189

15   The Calyx. (2024). Love your nature [Graphic panel at special exhibition]. Royal Botanic Garden, Sydney, Australia.

16   Darwin, C. (1859). *The Origin of Species*. Murray.

17   Odom, K. J., Hall, M. L., Riebel, K., Omland, K. E., & Langmore, N. E. (2014). Female song is widespread and ancestral in songbirds. *Nature Communications, 5*(1), 3379. doi.org/10.1038/ncomms4379

18   Haines, C. D., Rose, E. M., Odom, K. J., & Omland, K. E. (2020). The role of diversity in science: A case study of women advancing female birdsong research. *Animal Behaviour, 168*, 19–24. doi.org/10.1016/j.anbehav.2020.07.021

19   Santora, T. (2020). When female birds are overlooked, conservation suffers. *Audubon*. audubon.org/magazine/spring-2020/when-female-birds-are-overlooked-conservation

20   Alonso, N. (2021, February 12). Inside the movement to abolish colonialist bird names. *Outside*. outsideonline.com/culture/essays-culture/eponymous-bird-names-colonialism

21   Hazel, S. J., O'Dwyer, L., & Ryan, T. (2015). "Chickens are a lot smarter than I originally thought": Changes in student attitudes to chickens following a chicken training class. *Animals, 5*(3), 821–37. doi.org/10.3390/ani5030386

22   Marino, L. (2017). Thinking chickens: A review of cognition, emotion, and behavior in the domestic chicken. *Animal Cognition, 20*(2), 127–47. doi.org/10.1007/ s10071-016-1064-4

23 Riters, L. V., Kelm-Nelson, C. A., & Spool, J. A. (2019). Why do birds flock? A role for opioids in the reinforcement of gregarious social interactions. *Frontiers in Physiology, 10*, 421. doi.org/10.3389/fphys.2019.00421

24 University of Oxford. (2017, May 17). *Good grief! Losing a friend brings wild birds closer together.* University of Oxford News & Events. ox.ac.uk/news/2017-05-17-good-grief-losing-friend-brings-wild-birds-closer-together

25 Alwahaidi, K. (2023, April 25). *Parrots who learn to video call feathered friends feel less lonely, study finds.* CBC Radio. cbc.ca/radio/asithappens/parrots-who-learn-to-video-call-feathered-friends-feel-less-lonely-study-finds-1.6822228

26 Veit, W. (2020, March 13). 4 years of *Animal Sentience. Psychology Today.* psychologytoday.com/intl/blog/science-and-philosophy/202003/4-years-animal-sentience

27 Ahmed, A., & Corradi, C. (2022). *Animal sentience and consciousness: A review of current research.* Nuffield Council on Bioethics. academia.edu/100045906/animal_sentience_and_consciousness

28 Falk, D. (2024, April 19). Insects and other animals have consciousness, experts declare. *Quanta Magazine.* quantamagazine.org/insects-and-other-animals-have-consciousness-experts-declare-20240419

29 Cook, M. (2021, November 28). Octopuses, crabs and lobsters will be protected as sentient beings in UK. *BioEdge.* bioedge.org/bioethics-d75/animal-ethics-bioethics-d75/octopuses-crabs-and-lobsters-will-be-protected-as-sentient-beings-in-uk

30 Kohda, M., Bshary, R., Kubo, N., Awata, S., ... & Sogawa, S. (2023). Cleaner fish recognize self in a mirror via self-face recognition like humans. *Proceedings of the National Academy of Sciences, 120*(7), e2208420120. doi.org/10.1073/pnas.2208420120

31 Collett, R. (1911). *Norges pattedyr* [Norway's mammals]. Aschehoug, 362.

32 Baraniuk, C. (2018, February 9). Birds "dream sing" by moving their vocal muscles in their sleep. *New Scientist.* newscientist.com/article/2160842

33 Schnell, A. K., Clayton, N. S., Hanlon, R. T., & Jozet-Alves, C. (2021). Episodic-like memory is preserved with age in cuttlefish. *Proceedings of the Royal Society B: Biological Sciences, 288*(1957), 20211052. doi.org/10.1098/rspb.2021.1052

34 Galpayage Dona, H. S., Solvi, C., Kowalewska, A., Mäkelä, K., ... & Chittka, L. (2022). Do bumble bees play? *Animal Behaviour, 194*, 239–51. doi.org/10.1016/j.anbehav.2022.08.013

35 Bryn, B. (2014, June 12). *Science*: Crayfish can be calmed with anti-anxiety medication. American Association for the Advancement of Science. aaas.org/news/science-crayfish-can-be-calmed-anti-anxiety-medication

36 Cirelli, C., & Bushey, D. (2008). Sleep and wakefulness in *Drosophila melanogaster. Annals of the New York Academy of Sciences, 1129*(1), 323–29. doi.org/10.1196/annals.1417.017

37 University of Portsmouth. (2023, July 12). Belief in animals' capacity for emotion linked to better health and welfare. University of Portsmouth News. port.ac.uk/news-events-and-blogs/news/belief-in-animal-s-capacity-for-emotion-linked-to-better-health-and-welfare

38   Marino, L. (2019, May 8). Face it: A farmed animal is someone, not something. *Aeon*. aeon.co/essays/face-it-a-farmed-animal-is-someone-not-something

39   Kendrick, K. M., da Costa, A. P., Leigh, A. E., Hinton, M. R., & Peirce, J. W. (2001). Sheep don't forget a face. *Nature, 414*(6860), 165–66. doi.org/10.1038/35102669

40   Croney, C. C., & Reynnells, R. D. (2008). The ethics of semantics: Do we clarify or obfuscate reality to influence perceptions of farm animal production? *Poultry Science, 87*(2), 387–91. pubmed.ncbi.nlm.nih.gov/18212386

41   Benningstad, N. C. G., Kunst, J. R. (2020). Dissociating meat from its animal origins: A systematic literature review. *Appetite, 147*, 104554. doi.org/10.1016/j.appet.2019.104554

42   Zaraska, M. (2016, July 1). Meet the meat paradox. *Scientific American*. doi.org/10.1038/scientificamericanmind0716-50

43   Ritchie, H. (2023, September 25). *How many animals are factory-farmed?* Our World in Data. ourworldindata.org/how-many-animals-are-factory-farmed

44   Stucki, S. (2023). Animal warfare law and the need for an animal law of peace: A comparative reconstruction. *The American Journal of Comparative Law, 71*(1), 189–233. doi.org/10.1093/ajcl/avad018

45   Cowie, H. (2023, September 11). *In 200 years of animal welfare concerns, cruelty remains a significant issue, study shows*. University of York. york.ac.uk/news-and-events/news/2023/research/animal-cruelty-remains-a-significant-issue

46   Scott, G. R., Hawkes, L. A., Frappell, P. B., Butler, P. J., … & Milsom, W. K. (2015, March 1). How bar-headed geese fly over the Himalayas. *Physiology, 30*(2), 107–15. doi.org/10.1152/physiol.00050.2014

47   Grandin, T. (2013). Making slaughterhouses more humane for cattle, pigs, and sheep. *Annual Review of Animal Biosciences, 1*, 491–512. doi.org/10.1146/annurev-animal-031412-103713

48   PR Newswire. (2024, May 14). *New data shows 89% of cage-free egg commitments are fulfilled by food corporations*. Yahoo! Finance. ca.finance.yahoo.com/news/data-shows-89-cage-free-124200883.html

49   Torrella, K. (2021, March 23). The biggest animal welfare success of the past 6 years, in one chart. *Vox*. vox.com/future-perfect/22331708/eggs-cages-chickens-hens-meat-poultry

50   Block, K. (2024, March 1). *Major progress! 40% of hens used for eggs in the U.S. are now cage-free*. Humane World for animals. humaneworld.org/en/blog/40-percent-hens-eggs-cage-free-united-states

51   Ewing-Chow, D. (2024, March 31). Legal action: Commission challenged over failure to ban caged farming. *Forbes*. forbes.com/sites/daphneewingchow/2024/03/31/legal-action-commission-challenged-over-failure-to-ban-caged-farming

52   Ritchie, H. (2023, September 25). *Do better cages or cage-free environments really improve the lives of hens?* Our World in Data. ourworldindata.org/do-better-cages-or-cage-free-environments-really-improve-the-lives-of-hens

53   United Nations Environment Programme. (2022, July 28). *In historic move, UN declares healthy environment a human right*. UNEP. unep.org/news-and-stories/story/historic-move-un-declares-healthy-environment-human-right

54 Rodríguez-Garavito, C. (2022, October 28). *More than human rights: What can we learn from trees, animals, and fungi?* Open Global Rights. openglobalrights.org/more-than-human-rights-trees-animals-fungi

55 Kimmerer, R. W. & Whybrow, H. (2017, July 17). Robin Wall Kimmerer on the language of animacy. *Orion Magazine.* orionmagazine.org/article/robin-wall-kimmerer-language-animacy

56 Tiwari, S. (2024, April 19). *Africa is already leading the plant-based future.* Corporate Knights. corporateknights.com/category-food/africa-plant-based-future-afro-veganism

57 Tindall, R. (2024, November 30). *How big is the Chinese market for plant-based foods?* China-Britain Business Council FOCUS. focus.cbbc.org/how-big-is-chinas-market-for-plant-based-foods; IndustryARC. (n.d.). *Asia plant based food market—forecast (2025–2031).* Report code FBR 0533. industryarc.com/report/19784/asia-plant-based-food-market.html

58 Market Data Forecast (2024). *Latin America plant-based protein market research report.* Report ID 9472. marketdataforecast.com/market-reports/latin-america-plant-based-protein-market; Triton Market Research. (n.d.). *North America plant-based food and beverage market 2023–2030.* tritonmarketresearch.com/reports/north-america-plant-based-food-and-beverage-market

59 Holland, F. (2023, November 7). More than half of Europeans reducing meat consumption, research shows. *Just Food.* just-food.com/news/over-half-consumers-reducing-annual-meat-intake-in-europe-research-shows

60 IMARC Group. (2024, December 6). *Plant based food market size, share, demands, growth analysis & industry report 2025–2033.* openPR. openpr.com/news/3775438/plant-based-food-market-size-share-demands-growth-analysis

61 Tiwari, S. (2024, April 19). *Africa is already leading the plant-based future.* Corporate Knights.

62 Corichi, M. (2021, July 8). *Eight-in-ten Indians limit meat in their diets, and four-in-ten consider themselves vegetarian.* Pew Research Center. pewresearch.org/short-reads/2021/07/08/eight-in-ten-indians-limit-meat-in-their-diets-and-four-in-ten-consider-themselves-vegetarian

63 Xu, X., Sharma, P., Shu, S., Lin, T., ... & Jain, A. K. (2021). Global greenhouse gas emissions from animal-based foods are twice those of plant-based foods. *Nature Food, 2,* 724–32. doi.org/10.1038/s43016-021-00358-x

64 Hunter, L. (2024, June 25). Denmark announces world-first climate tax on agriculture—earmarks billions for rewilding. *The Copenhagen Post.* cphpost.dk/2024-06-25/news/climate/denmark-announces-world-first-climate-tax-on-agriculture-earmarks-billions-for-rewilding

65 DePape, K. (2024, December 5). *Small changes can cut your diet-related carbon footprint by 25%.* McGill University Newsroom. mcgill.ca/newsroom/channels/news/small-changes-can-cut-your-diet-related-carbon-footprint-25-355698

66 Ministry of Food, Agriculture and Fisheries of Denmark. (2023). *Danish Action Plan for Plant-Based Foods.* en.fvm.dk/news-and-contact/focus-on/action-plan-on-plant-based-foods

67   De Lorenzo, D. (2023, November 23). How Denmark made the plant-based action plan possible. *Forbes*. forbes.com/sites/danieladelorenzo/2023/11/23/how-denmark-made-the-plant-based-action-plan-possible

68   United States Conference of Mayors. (2023). *2023 adopted resolutions.* legacy.usmayors.org/resolutions/91st_conference/proposed-review-list-full-print-committee-all.asp; c40 Cities, Arup, & University of Leeds. (2019). *Addressing food-related consumption-based emissions in c40 cities.* c40 Knowledge Hub. c40knowledgehub.org/s/article/in-focus-addressing-food-related-consumption-based-emissions-in-c40-cities

69   NYC Health + Hospitals. (2024, March 14). *NYC Health + Hospitals celebrates 1.2 million plant-based meals served.* NYC Health + Hospitals, press release #038-24. nychealthandhospitals.org/pressrelease/nyc-health-hospitals-celebrates-1-2-million-plant-based-meals-served

70   Fang, S., & Li, X. (2020). Historical ownership and territorial disputes. *The Journal of Politics, 82*(1), 345–60. doi.org/10.1086/706047

71   Dupuis-Désormeaux, M., Dheer, A., Gilisho, S., Kaaria, T. N., … & MacDonald, S. E. (2023). Teeth, tusks, and spikes: Repeated den sharing between predator and prey in an African savannah. *African Journal of Ecology, 61*(4), 1006–9. doi.org/10.1111/aje.13153

72   Fernandez, V., Abdala, F., Carlson, K. J., Collins Cook, D., … & Tafforeau, P. (2013). Synchrotron reveals early triassic odd couple: Injured amphibian and aestivating therapsid share burrow. *PLoS One, 8*(6), e64978. doi.org/10.1371/journal.pone.0064978; Kondo, A. (2018). Interspecific burrow sharing between mammals in countryside in Japan. *Mammal Study, 43*(3). doi.org/10.3106/ms2017-0062; Coppola, F., Dari, C., Vecchio, G., Scarselli, D., & Felicioli, A. Cohabitation of settlements among crested porcupine (*Hystrix cristata*), red fox (*Vulpes vulpes*) and european badger (*Meles meles*). *Current Science 119*(5). doi.org/10.18520/cs/v119/i5/817-822

73   Parks Canada. (2014, February 13). A wild way to move—Banff National Park [Video]. YouTube. youtube.com/watch?v=9JX6cqME6HW

74   Peninsula Open Space Trust. (2020). Coyote and badger playing together [Video]. YouTube. youtube.com/watch?v=2bICTWNRrGE

75   ojalehto, b. l., Medin, D. L., Horton, W. S., Garcia, S. G., & Kays, E. G. (2015). Seeing cooperation or competition: Ecological interactions in cultural perspectives. *Topics in Cognitive Science, 7*(4), 624–45. doi.org/10.1111/tops.12156

76   Massarella, K., Krauss, J. E., Kiwango, W., & Fletcher, R. (2022). Exploring convivial conservation in theory and practice: Possibilities and challenges for a transformative approach to biodiversity conservation. *Conservation and Society, 20*(2), 59–68. doi.org/10.4103/cs.cs_53_22

77   WWF Tigers Alive. (2023). *Living with tigers in a fast changing world.* tigers.panda.org/news_and_stories/stories/living_with_tigers_in_a_fast_changing_world

78   Carrington, D. (2024, January 5). African elephant populations stabilise in southern heartlands. *The Guardian*. theguardian.com/environment/2024/jan/05/african-elephant-populations-stabilise-in-southern-heartlands

79 Rex, P. T., May, J. H., III, Pierce, E. K., & Lowe, C. G. (2023). Patterns of overlapping habitat use of juvenile white shark and human recreational water users along southern California beaches. *PLoS One, 18*(6), e0286575. doi.org/10.1371/journal.pone.0286575

80 Freedman, E. (2023, June 9). Great white sharks have almost no interest in eating humans, study confirms. *Live Science.* livescience.com/animals/sharks/great-white-sharks-have-almost-no-interest-in-eating-humans-study-confirms

81 Martins, R. (2015, February 24). Rats remember who's nice to them—and return the favor. *National Geographic.* nationalgeographic.com/animals/article/150224-rats-helping-social-behavior-science-animals-cooperation

82 Dean, K. R., Krauer, F., Walløe, L., Lingjærde, O. C., ... & Schmid, B. V. (2018). Human ectoparasites and the spread of plague in Europe during the Second Pandemic. *Proceedings of the National Academy of Sciences, 115*(6), 1304–9. doi.org/10.1073/pnas.1715640115

83 Albery, G. F., Carlson, C. J., Cohen, L. E., Eskew, E. A., ... & Becker, D. J. (2022). Urban-adapted mammal species have more known pathogens. *Nature Ecology & Evolution, 6*(6), 794–801. doi.org/10.1038/s41559-022-01723-0

84 Camp, J. V., Desvars-Larrive, A., Nowotny, N., & Walzer, C. (2022). Monitoring urban zoonotic virus activity: Are city rats a promising surveillance tool for emerging viruses? *Viruses, 14*(7), 1516. doi.org/10.3390/v14071516

85 Scheggia, D., & Papaleo, F. (2020). Social neuroscience: Rats can be considerate to others. *Current Biology, 30*(6), R274–76. doi.org/10.1016/j.cub.2020.01.093

86 Peterson, C. (2024, April 30). *New regulations look to save the lives of Missoula's bears.* KPAX News. kpax.com/news/missoula-county/new-regulations-look-to-save-the-lives-of-missoulas-bears

87 Marshall-Pescini, S., Schwarz, J. F. L., Kostelnik, I., Virányi, Z., & Range, F. (2017). Importance of a species' socioecology: Wolves outperform dogs in a conspecific cooperation task. *Proceedings of the National Academy of Sciences, 114*(44), 11793–98. doi.org/10.1073/pnas.1709027114

88 Dwyer, R. J., Brady, W. J., Anderson, C., & Dunn, E. W. (2023). Are people generous when the financial stakes are high? *Psychological Science, 34*(9), 999–1006. doi.org/10.1177/09567976231184887

89 University of California, Los Angeles. (2023, April 23). Small acts of kindness are frequent and universal, study finds. *ScienceDaily.* sciencedaily.com/releases/2023/04/230424162911.htm

90 Wadley, J. (2020). *Children show altruism at a young age.* Michigan Today, University of Michigan. michigantoday.umich.edu/2020/07/23/children-show-altruism-at-a-young-age

91 Keltner, D. (2023, January 24). What's the most common source of awe? *Greater Good Magazine.* greatergood.berkeley.edu/article/item/whats_the_most_common_source_of_awe

92 James, I. (2023, October 6). The largest dam removal in history stirs hopes of restoring California tribes' way of life. *Los Angeles Times.* latimes.com/environment/story/2023-10-05/klamath-dam-removal-tribes

93 Kimbrough, L. (2024, October 24). *Largest dam removal ever, driven by tribes, kicks off Klamath River recovery.* Mongabay. news.mongabay.com/2024/10/largest-dam-removal-ever-driven-by-tribes-kicks-off-klamath-river-recovery

94 Sherriff, L. (2024, November 25). *After 100 years, salmon have returned to the Klamath River—following a historic dam removal project in California.* BBC. bbc.com/future/article/20241122-salmon-return-to-californias-klamath-river-after-dam-removal

95 Eve Tuck, evetuck.com

96 Griffiths, T. (2022, December 13). Australia burning. *Springs: The Rachel Carson Center Review*, 2. doi.org/10.5282/rcc-springs-2856

97 Jolly, C. J., & Nimmo, D. (2022, January 6). Surprisingly few animals die in wildfires—and that means we can help more in the aftermath. *The Conversation.* theconversation.com/surprisingly-few-animals-die-in-wildfires-and-that-means-we-can-help-more-in-the-aftermath-174392; Jolly, C. J., Dickman, C. R., Doherty, T. S., van Eeden, L. M., ... & Nimmo, D. G. (2022). Animal mortality during fire. *Global Change Biology*, 28(6), 2053–65. doi.org/10.1111/gcb.16044

98 McCoy, M. K. (2023, May 11). *After Australia's bushfires, AI cameras capture wildlife recovery.* Conservation International. conservation.org/blog/after-australias-bushfires-ai-cameras-capture-wildlife-recovery

99 Moore, G. (2024, February 21). *Hard to kill: Here's why eucalypts are survival experts.* Find an Expert: The University of Melbourne. findanexpert.unimelb.edu.au/news/77668

100 Gorta, S. B. Z., Callaghan, C. T., Samonte, F., Ooi, M. K. J., ... & Cornwell, W. K. (2023). Multi-taxon biodiversity responses to the 2019–2020 Australian megafires. *Global Change Biology*, 29(23), 6727–40. doi.org/10.1111/gcb.16955

101 Cruickshank, A. (2023, June 21). Vancouver's development destroyed Burrard Inlet. Tsleil-Waututh Nation is determined to save it. *The Narwhal.* thenarwhal.ca/burrard-inlet-vancouver-tsleil-waututh

102 Ore, J. (2022, November 4). *Aysanabee's pandemic phone calls with his grandfather inspired his debut album.* CBC. cbc.ca/radio/unreserved/aysanabee-album-watin-grandfather-1.6639855

103 Ritchie, H. (2022, November 30). *How many species are there?* Our World in Data. ourworldindata.org/how-many-species-are-there

104 Pearce, F. (2015, August 17). *Global extinction rates: Why do estimates vary so wildly?* Yale Environment 360. e360.yale.edu/features/global_extinction_rates_why_do_estimates_vary_so_wildly

105 Ashworth, J. (2024, March 14). *Number of threatened bird species falls in latest conservation update.* Natural History Museum. nhm.ac.uk/discover/news/2024/march/number-threatened-bird-species-falls-latest-conservation-update.html

106 Williams, R., Lacy, R. C., Ashe, E., Barrett-Lennard, L., ... & Paquet, P. (2024). Warning sign of an accelerating decline in critically endangered killer whales (*Orcinus orca*). *Communications Earth & Environment*, 5(1), 173. doi.org/10.1038/s43247-024-01327-5

107 The Whale Museum. (n.d.). *FAQ about the Southern Resident endangered orcas.* whalemuseum.org/pages/frequently-asked-questions-about-the-southern-resident-endangered-orcas

108 Keenan, G. (2016, September 21). Fish recorded singing dawn chorus on reefs just like birds. *New Scientist.* newscientist.com/article/2106331-fish-recorded-singing-dawn-chorus-on-reefs-just-like-birds

109 Manning, J. (2024, January 18). *Scientists study how underwater soundscapes and young fish could help the Reef.* Australia Institute of Marine Science. aims.gov.au/information-centre/news-and-stories/scientists-study-how-underwater-soundscapes-and-young-fish-could-help-reef

110 Aoki, N., Weiss, B., Jézéquel, Y., Zhang, W. G., ... & Mooney, T. A. (2024). Soundscape enrichment increases larval settlement rates for the brooding coral *Porites astreoides.* *Royal Society Open Science, 11*(3), 23514. doi.org/10.1098/rsos.231514

111 Lamont, T. A. C., Williams, B., Chapuis, L., Prasetya, M. E., ... & Simpson, S. D. (2021). The sound of recovery: Coral reef restoration success is detectable in the soundscape. *Journal of Applied Ecology, 59*(3), 742–56. doi.org/10.1111/1365-2664.14089

112 Lambert, J. (2021, December 9). Cleared tropical forests can regain ground surprisingly fast. *Science News.* sciencenews.org/article/tropical-forest-recovery-ecosystem

113 Quote from Poorter, L. (2021, December 9). *Tropical forests regrow surprisingly fast.* Wageningen University & Research. wur.nl/en/research-results/research-institutes/environmental-research/show-wenr/tropical-forests-regrow-surprisingly-fast.htm; Poorter, L., Craven, D., Jakovac, C. C., Van Der Sande, ... & Hérault, B. (2021). Multi-dimensional tropical forest recovery. *Science, 374*(6573), 1370–76. doi.org/10.1126/science.abh3629

114 Field Museum. (n.d.). *Monarch community science project.* fieldmuseum.org/activity/monarch-community-science-project

115 Wilson, J. (2024, May 4). Sebastião Salgado [Audio podcast episode]. In *This Cultural Life.* BBC Radio 4. bbc.co.uk/programmes/m001yqp9

116 Locatelli, A. (2022, October 17). Meet the adventurer: Photographer Sebastião Salgado on reviving the forests of Brazil's Vale do Rio Doce. *National Geographic.* nationalgeographic.com/travel/article/meet-adventurer-photographer-sebastio-salgado-reviving-forests-brazil-vale-do-rio-doce

117 Wollowski, C. (2024, September 7). *"Planting trees to be able to live in peace."* Deutschland.de. deutschland.de/en/topic/environment/sebastiao-salgado-nature-conservation-in-brazil

118 Klinger, K. R., Hasle, A. F., & Oberhauser, K. S. (2024). Characteristics of urban milk-weed gardens that influence monarch butterfly egg abundance. *Frontiers in Ecology and Evolution, 12.* doi.org/10.3389/fevo.2024.1444460

119 Wu, S., Chen, B., Webster, C., Xu, B., & Gong, P. (2023). Improved human greenspace exposure equality during 21st century urbanization. *Nature Communications, 14*(1), 6460. doi.org/10.1038/s41467-023-41620-z

120 Filippi, P., Congdon, J. V., Hoang, J., Bowling, D. L., ... & Güntürkün, O. (2017). Humans recognize emotional arousal in vocalizations across all classes of terrestrial vertebrates: Evidence for acoustic universals. *Proceedings of the Royal Society B: Biological Sciences, 284*(1859), 20170990. doi.org/10.1098/rspb.2017.0990

121 Bosman, A. (2022, December 21). *Empathetic humans are better at understanding animal sounds.* Earth.com. earth.com/news/empathetic-humans-are-better-at-understanding-animal-sounds

122 TED Talks. (2016, February). This is your brain on communication | Uri Hasson | TED2016 [Video]. ted.com/talks/uri_hasson_this_is_your_brain_on_communication

123 Princeton University. (2020, January 20). Baby and adult brains "sync up" during play. *ScienceDaily*. sciencedaily.com/releases/2020/01/200109163956.htm

124 Wallauer, B. (2017, June 6). *Exploring evolution and spirituality in chimpanzees and humans*. Jane Goodall's Good for All News. news.janegoodall.org/2017/06/06/evolution-spirituality-and-humanity

125 Pryke, L. (2023, July 13). Friday essay: From angry gods and fertile myths to battleships and new technologies—how the wind shapes our world. *The Conversation*. theconversation.com/friday-essay-from-angry-gods-and-fertile-myths-to-battleships-and-new-technologies-how-the-wind-shapes-our-world-206592

126 Sullivan, S., & Ganuses, W. S. (2020). Understanding Damara / ≠Nūkhoen and ǁUbun indigeneity and marginalisation in Namibia. In W. Odendaal & W. Werner (Eds.), *"Neither here nor there": Indigeneity, marginalisation and land rights in post-independence Namibia* (pp. 283–324). Land, Environment and Development Project, Legal Assistance Centre.

127 Schnegg, M. (2019). The life of winds: Knowing the Namibian weather from someplace and from noplace. *American Anthropologist, 121*(4), 830–44. doi.org/10.1111/aman.13274

128 Breyer, T. (n.d.). Bridging worlds: The synergy of philosophy and anthropology in the quest for a critical understanding of cultural meanings. *HIAS, 1*, Globalized Nature. Hamburg Institute for Advanced Study. hias-hamburg.de/hias-magazin/bridging-worlds

129 Villazon, L. (2023, July 12). How many trees does it take to produce oxygen for one person? *BBC Science Focus Magazine*. sciencefocus.com/planet-earth/how-many-trees-does-it-take-to-produce-oxygen-for-one-person

Special Appendix for Educators

1 Hurless, N., & Kong, N. Y. (2024). Strategies for delivering trauma-informed climate change instruction. *The Journal of Environmental Education, 56*(1), 36–48. doi.org/10.1080/00958964.2024.2423298

2 Ympäristöahdistus.fi. (n.d.). *Etusivu*. ymparistoahdistus.fi

3 See Pierre-Louis, K. (2023, May 1). Better bus systems could slow climate change. *Scientific American*. doi.org/10.1038/scientificamerican0523-74

Resources

The Climate Emotions Wheel, created by Anya Kamenetz and the Climate Mental Health Network in collaboration with Panu Pihkala, can be found in over thirty languages and various formats at climatementalhealth.net/wheel.

The website for *The Essential Toolkit for Climate Justice Educators* offers a wealth of activities and other resources at existentialtoolkit.com/more-tools.

The Global Investigative Journalism Network's "Reporting Guide to Holding Governments Accountable for Climate Change Pledges" is available at gijn.org/resource/guide-to-holding-governments-accountable-for-climate-change-pledges.

Here are some other resources mentioned in the book that you may find helpful as you develop and nurture your practice of hope.

- The Solutions Journalism Network: solutionsjournalism.org

- Project Drawdown: drawdown.org

- Gapminder: gapminder.org

- More Than Human Life: mothrights.org

- Global Optimism: globaloptimism.com/why-stubborn-optimism

- The Process of Eco-Anxiety: ecoanxietyprocess.com

DAVID
  SUZUKI
INSTITUTE

THE DAVID SUZUKI INSTITUTE is a companion organization to the David Suzuki Foundation, with a focus on promoting and publishing on important environmental issues in partnership with Greystone Books.

We invite you to support the activities of the Institute. For more information, please write to us at

info@davidsuzukiinstitute.org.